HOW TO ESTIMATE WITH METRIC UNITS

- **For Government Contracts**
- **For Foreign Construction Projects**

HOW TO ESTIMATE WITH METRIC UNITS

- **For Government Contracts**
- **For Foreign Construction Projects**

R.S. Means Company, Inc.
Southam Construction Information Network
CONSTRUCTION PUBLISHERS & CONSULTANTS
100 Construction Plaza, P.O. Box 800
Kingston, MA 02364-0800
(617) 585-7880

©1993

In keeping with the general policy of R.S. Means Company, Inc., its authors, editors, and engineers apply diligence and judgment in locating and using reliable sources for the information published. However, no guarantee or warranty can be given, and all responsibility and liability for loss or damage are hereby disclaimed by the authors, editors, engineers and publisher of this publication with respect to the accuracy, correctness, value and sufficiency of the data, methods, and other information contained herein as applied for any particular purpose or use.

The editors for this book were Mary Greene, Suzanne Morris, John Chiang, Roger Grant, Neil Smit, and John G. McConville. Composition was supervised by Joan C. Marshman. The book and jacket were designed by Norman R. Forgit.

No part of this publication may be reproduced, stored in a retrieval system, or transmitted in any form or by any means without prior written permission of R.S. Means Company, Inc.

Printed in the United States of America

10 9 8 7 6 5 4 3 2 1

Library of Congress Cataloging in Publication Data

ISBN 0-87629-317-8

Table of Contents

Acknowledgments	vi
Foreword	vii
Introduction: Opportunities and Challenges	ix
Part I: Metric Estimating Guidelines	1
Section 1: What Is SI Metric?	5
Section 2: Estimate Considerations	17
Section 3: Using *Means Building Construction Cost Data, Metric Edition*	29
Part II: Metric Estimating by MasterFormat Division	37
Division 1: General Conditions	41
Division 2: Site Work	47
Division 3: Concrete	55
Division 4: Masonry	69
Division 5: Metals	79
Division 6: Wood and Plastics	83
Division 7: Thermal and Moisture Protection	89
Division 8: Door and Windows	95
Division 9: Finishes	101
Division 10: Specialties	107
Division 11: Architectural Equipment	109
Division 12: Furnishings	111
Division 13: Special Construction	113
Division 14: Conveying Systems	115
Division 15: Mechanical	117
Division 16: Electrical	125
Part III: Metric in Design	131
Appendix A: Metric Conversion Tables	145
Appendix B: Metric Product Sources	155
Appendix C: Metric References	163
Appendix D: Professional Associations	169
Glossary of Terms	177
Index	179

Acknowledgments

Some of the material included in this publication was adapted from the *Metric Guide for Federal Construction*, ©1991, National Institute of Building Sciences, Washington, DC. Members of the Building Sciences' Construction Metrication Council were also very helpful in providing current information on the progress and effects of metrication on federal construction projects, and on sources of building products manufactured in metric dimensions.

We are grateful for the input of the many building product associations who provided up-to-date information on metrication of their respective products.

Lastly, our thanks to Wayne Watson of W^2 Consultants, Ltd. in Sherwood Park, Alberta, Canada, who provided the metric building plans shown in Part III.

Foreword

We must move to the metric system of measurement as quickly as possible! The Metric Conversion Act of 1975, the Omnibus Trade and Competitiveness Act of 1988 and Executive Order 12770 of July 25, 1991 require the Federal Government to convert to metric and be the leader in moving the private sector to metric. While English is the world language for oral and written communication, certainly metric is the world system for communication of measurement.

At the end of World War II the United States of America produced 75% of the world's products and services. Today we produce only 25% of all the goods and services produced in the world. We have fallen from the position of primary producer of all goods and services to just one of the world's mighty, along with Germany, Japan, France, Great Britain, Canada, Russia and China, among others.

International trade is the opportunity to develop wealth, create jobs, and in general advance the quality of life in America. Look around you at the products produced abroad, including VCRs, televisions, automobiles, and clothing. We must ensure that the world at large is getting at least as many of their products from American producers. America was once able to dictate product lines and services because we were so far advanced technologically beyond the rest of the world. Today the area where we show notable advantage is in the art of war! Obviously we cannot, nor do we want to, depend on war to be economically successful.

So how can America succeed in the international trade arena? We must be the most productive, the most competitive, and produce the highest quality products. The metric system of measurement is a critical element. Using a base of 10 versus the base of 12 in the English system, all transactions are much simpler, more logical and consistent. In addition, metric is the world communication standard, and we simply will not be able to sell if the customer cannot easily understand or apply the measurement of our products.

The General Services Administration has made outstanding progress and will procure/produce all buildings/facilities in metric beginning January 1994. The Department of Defense Construction Program is moving rapidly to metric, with the Army Corps of Engineers leading the way. The Federal Highway Administration will issue all Federal Highway funds for metric construction in 1996. The other federal agencies are also committing to metrication in the next five years.

I urge all elements of the architectural, engineering and construction industry to "step up to the plate" and move to metric as quickly as possible. Since we are going to "soft" metric first, with a gradual transition to hard metric and product sizes only as industry agrees, there will be very little additional expense, with a tremendous potential for long-range benefits.

Thomas R. Rutherford, P.E., Chairman
Federal Metrication Construction Committee

Introduction: Opportunities and Challenges

Of the approximately 180 countries that are presently represented in the United Nations, only three use the old English Imperial measurement system (feet, pounds, cubic yards, and so on). These three countries are Burma (now called Myanmar), Liberia, and the United States. The other 177 countries use the System International (SI) metric system of measurement. The movement to the metric system in the United States has been one of fits and starts. It has been discussed, legislated, forgotten, and discussed again for the better part of a century. The United States is the last major nation to convert to metric. Because American products must be able to compete in foreign metric markets, adopting the metric system is crucial now and in the future for the acceptance of U.S. products and services in overseas markets.

The Metric Conversion Act of 1975, as amended by the Omnibus Trade and Competitiveness Act of 1988, establishes that, to the extent feasible, the metric system be used in all federal government procurements, grants, and other business-related activities by September 30, 1992. Federal agencies involved in construction have agreed to institute the use of metric in the design of all federal construction by January, 1994. GSA expects that its construction projects will all be metric by 1995, with smaller projects leading the way. The Army Corps of Engineers plans to have over $650 million in metric construction under way by 1995. The Department of State is already doing all of its work in metric, and the Treasury Department expects all of its construction to be in metric by 1994. Other federal agencies have similar timetables. U.S. Government construction represents approximately $40 billion annually of the $450 billion U.S. construction industry. By the turn of the century, it is anticipated that the U.S. will be fully metricated.

This book is intended to help the construction professional meet the challenge of metrication. There has been much speculation about the difficulty of converting to metric in the U.S. construction industry. Metric conversion is not necessarily a complicated process; it will take time and careful planning to make the transition as efficient as possible. This publication is intended to alleviate some fears and concerns, and prepare readers to think and estimate in metric. It introduces the basics of the metric system, including reading and writing parameters, where to find hard metric products (that is, those produced in rational or convenient metric sizes), and the implications of "metric construction" on productivity.

The use of decimal arithmetic and simple base units make the system convenient and quite easy to learn and use. Although most nations have been using the metric system of measurement for centuries, several countries including Great Britain, Canada, and South Africa have converted in the past 20 years. The experiences of construction professionals in these nations indicate that, despite initial concerns, the changeover was relatively painless in terms of the cost of construction and the adjustment of the personnel involved in the construction process. Figure i.1 summarizes the metric conversion process in these countries.

Metrication of the construction industry in the U.S. is taking place now because of the lead role being played by the federal government. In the past, the construction industry was uncertain of the federal commitment, and had no apparent economic incentive to convert.

For those sectors of the U.S. construction industry that export goods or services, metrication is vital for these reasons:

- In 1990, U.S. nonlumber construction product exports totaled about $2.8 billion, and imports totaled about $4.2 billion.
- Foreign billings for American architecture/engineering/contracting firms amounted to $3.2 billion in 1989, about a third of which was from Europe.
- The European community, now the world's largest market, has specified that products with nonmetric labels will not be permitted for sale after 1992.
- The largest U.S. trading partners, Canada and Mexico, are now predominantly metric countries.
- During the U.S.-Japanese Strategic Impediments Initiative negotiations of the early 1990s, the Japanese identified nonmetric U.S. products as a specific barrier to the importation of U.S. goods into Japan.
- The failure of U.S. companies to manufacture and sell metric products overseas represents billions of dollars in lost sales annually.

The modern metric system was established by international agreement in 1960. It now is the standard international language of measurement and the system mandated by the *Metric Conversion Act* for use in the United States.

The metric system is coherent – only one unit is used for each physical quantity, and there are no conversion factors or constants to remember. For example, the meter (and its decimal multiples) is the single metric measure for length, while its inch-pound system equivalents include the mil, the inch (1000 mils), the foot (12"), the yard (3'), the fathom (6'), the rod (16.5'), the chain (66'), the furlong (660'), and the mile (5,280').

Let's look at some of the major benefits derived from the upcoming change to the metric system. First, use of international units improves communication in international trade and results in increased sales and lower manufacturing and distribution costs. Finding an overseas vendor for parts is easier, and communication is less prone to misunderstanding. For example, the United States lost several hundred million dollars in potential lumber sales to North Africa and the Middle East in the 1980s because products could not be provided in metric sizes. Since then, U.S. mills, sawmills, and lumber companies have substantially increased the export market for their products by adopting metric dimensions.

Second, making the transition to metric offers a one-time chance to rationalize sizes, standardize equipment elements, and modularize product packaging. With fewer sizes and pieces, inventory is reduced and its control is simpler. The result is reduced production cost.

International Precedent in Metrication in Construction — Comparison of Program Timing

Country	Metric Planning or Coordinating Committee (Date Established)	First Year Shown in the Official Metric Program	M-Day on Start of Metric Construction	Time to M-Day (Years)	Program Completion Date	Total Conversion Period	Remarks
Britain	Construction Industry Metric Panel (British Standards Institution), May 1965 Steering Committee for Industrial Materials and Construction (Metrication Board), June 1969	1966 Program Available February 1967	From January 1970 (No M-Day)	4 Years	Estimated: End of 1972 Actual: End of 1974	7 Years (9 Years)	Conversion was hindered by a lack of organization and an absence of metric conversion activity in the general economy. Metrication was combined with dimensional coordination.
South Africa	Industrial Metrication Division, Metrication Department (South African Bureau of Standards), during 1968. Metrication Advisory Board established 1967	1969	From January 1, 1972 (Subsequently advanced to July 1, 1971) (No M-Day)	3 Years (2-1/2 Years)	End of 1973	5 Years	Metric implementation took place simultaneously throughout the entire economy. Target for metric construction was advanced by 6 months to sustain momentum. Dimensional coordination was not mandated.
Australia	Building and Construction Advisory Committee (Metric Conversion Board), November 1970	1971	January 1, 1974	3 Years	End of 1976	6 Years	Program completed within the assigned time frame. Importance of commitment by government sector, and simultaneous metrication throughout the economy.
Canada	Steering Committee 5 (Construction Industry) (Metric Commission Canada) mid-1972 Sector Committee 5.1 (Construction – including design) First Meeting, August 1973	1975	January 1, 1978 (1978 Designated as M-Year)	3 Years	End of 1980	6 Years	Program predominantly on schedule, with slippage in some provinces. Metric leadership by the government sector. M-Day twice deferred during planning stage.

Figure i.1

Third, design and quality control in metric measurements and standards have been reported to be quicker and involve fewer errors, resulting in reduced cost and improved quality. Training and retraining of personnel is less problematic. Also, a study reported by the Department of Defense shows that engineering productivity improved by 5% with the use of metric units.

A survey published by the Society for the Advancement of Management reported that 16% of Fortune 1000 companies lost sales because they could not or would not supply in metric. A Department of Commerce memorandum issued in the mid-1980s stated that U.S. machine tools "are not metric and, therefore, increasingly unsaleable... Europeans are no longer willing to accept (soft conversions and nonmetric products)..." Major European companies such as VW and Daimler no longer invite U.S. firms to bid for tooling contracts.

A major U.S. appliance manufacturer reported in 1980 that "We do not do well in Europe from a U.S. export viewpoint (because of) configuration differences... The world is currently trending to European designs... (lack of) metrication does inhibit sales. By its reluctance to (join other nations in using the metric system), the U.S. is losing billions of dollars in sales yearly."

The following lists summarize some of the typical benefits and penalties associated with the upcoming conversion of the U.S. construction industry to the metric system of measurement.

Metrication: Benefits

Variety Reduction
- Rationalization of product line
- Production of preferred products
- Establishment of a more logical size range for new markets
- Standardization of materials and equipment

Development of New Markets
- Expansion of sales into new areas
- Export of skill

Increases in Productivity
- Simplicity of decimal measurement
- Quicker, more accurate calculations
- Greater accuracy in measurement and production (fewer errors)

Review and Rationalization
- Simplification of administrative and technical procedures/details
- Rationalization of organizational practices and work processes
- Harmonization of differing approaches (specifications, standards, codes)

Data Improvement and Revision
- Opportunity for data review and improvement
- Up-to-date information or technology
- Worldwide exchange of technical data

More Efficient Production
- Redesign of old products/techniques
- Use of planned obsolescence to acquire efficient equipment
- Better product quality and longer production runs

Better Communication
- Better informed staff (updating of skills)

- Improved communications within the organization and with suppliers, clients or customers

Metrication: Penalties

Duplication
- Dual equipment and material requirement (existing and older facilities)
- Dual product line
- Dual inventories and records
- Return of incorrect orders

Loss of Markets
- Competition from metric countries
- Competition from specialist manufacturers

Loss of Productivity
- Work slow-down during transition
- Work slow-down through unfamiliarity
- Errors
- Loss of staff time in training

Metric Program Costs
- Metric program manager
- Additional support staff and facilities
- Time lost in meetings/conferences

Development/Acquisition of New Data
- Need for new technical literature
- Need for new reference data, forms, charts, guidelines, etc.

Costs of New or Modified Equipment
- New tools and measuring instruments
- Modification or recalibration of equipment
- Incremental equipment costs

Training and Familiarization Costs
- Cost of training program
- Cost of familiarization of suppliers and customers

Support for Metrication

Several factors should make the conversion to metric construction easier in the U.S. These include:

- Inch-pound and metric units are already included in two of the three model building code standards (BOCA and SBCCI).
- Computerized design and estimating functions, as well as computer-controlled HVAC and manufacturing systems, can simplify metric conversion.
- U.S. design and construction firms currently use metric in foreign construction work, without problems.
- Early GSA construction projects in metric have shown no increase in costs related to design or construction.
- The cost of converting to metric in other American industries has been far lower than expected, with significantly greater benefits than expected.

Metrication should be viewed enthusiastically as an opportunity for planned change — a national "common cause." The goal is to achieve the benefits inherent in a more rational measurement system that is in step with the rest of the world community.

Estimating construction costs, regardless of the measurement system, requires a well-rounded, working knowledge of how buildings are constructed. The successful estimator is one who has taken the opportunity to learn as much as possible about the building and the measurement system being used. The single most important credential that the estimator takes to the job is experience. Without the ability to visualize the construction process in the mind's eye, the estimator surely cannot do the task properly. A second desirable trait is the ability to see a new way to complete a particular work activity, using a new piece of equipment or a new technology. Still, there are times when an estimator must go beyond his or her own experience, using resources such as this book and often working directly with experienced construction superintendents and engineers to successfully accomplish the construction cost estimate.

Part I
METRIC ESTIMATING GUIDELINES

liter
kg
mm
joule
pascal
tonne
Celsius
m³

Part I
Metric Estimating Guidelines

The following sections contain guidelines for taking quantities off in metric dimensions for an estimate. Section 1 defines SI Metric base units and prefixes, and outlines the rules for writing metric names, and for converting and rounding. Charts in Section 1 list the key metric units used in various types of construction, by trade. Section 2 describes some special considerations of metric in the context of general estimating procedures. Section 3, *"Using Means Building Construction Cost Data, Metric Version,"* explains how this cost reference was developed, how it is organized, and how to use the data effectively. A CSI division-by-division breakdown on metric estimating follows in Part II.

Section 1
What Is SI Metric?

System International (SI) Metric is the name given to the modern measurement system that has been adopted as a worldwide standard for the preferred system of weights and measures.

In the SI Metric system, only one unit is used to describe a measured quantity. It is infinitely simple because there are no conversions to remember (inches, feet, yards, fathoms, rods, chains, furlongs, miles, etc.). All quantities are derived using decimal arithmetic from the base units. Listed below are the base units used in construction.

Quantity or Measurement	SI Unit	Symbol
Length	meter	m
Mass (Weight)	kilogram	kg
Time	second	s
Electric current	ampere	A
Temperature	kelvin	K
Luminous intensity	candela	cd

All other SI units are derived from these base units.

It is important to note that in using symbols for SI units, the proper upper case or lower case letters must be used.

Listed below are the SI decimal prefixes used in construction. The most commonly used prefixes in construction are kilo (k) and milli (m).

Factor	Prefix	Symbol
$1\ 000\ 000\ 000 = 10^9$	giga	G
$1\ 000\ 000 = 10^6$	mega	M
$1\ 000 = 10^3$	kilo	k
$100 = 10^2$	hecto	h
$10 = 10^1$	deka	da
$0.1 = 10^{-1}$	deci	d
$0.01 = 10^{-2}$	centi	c
$0.001 = 10^{-3}$	milli	m

Writing Metric Symbols and Names

- Print unit symbols in upright type in lower case, except for liter (L) or unless the unit name is derived from a proper name.
- Print unit names in lower case, even those derived from a proper name.
- Print decimal prefixes in lower case for magnitudes 10^3 and lower (that is, k, m, and n) and print the prefixes in upper case for magnitudes 10^6 and higher (that is, M and G).
- Leave a space between a numeral and a symbol (write 45 kg or 37 °C, not 45kg or 37°C).
- Do not use a degree mark (°) with kelvin temperature; write K, not °K.
- Do not leave a space between a unit symbol and its decimal prefix (write kg, not k g).
- Do not use the plural of unit symbols (write 45 kg, not 45 kgs), but do use the plural of written unit names (several kilograms).
- For technical writing, use symbols in conjunction with numerals (the area is 10 m^2); write out unit names if numerals are not used (carpet is measured in square meters). Numerals may be combined with written unit names in nontechnical writing (10 meters).
- Indicate the product of two or more units in symbolic form by using a dot positioned above the line (Kg·m).
- Do not mix names and symbols (write N·m or newton meter, not N·meter or newton·m).
- Do not use a period after a symbol (write 12 g, not 12 g.), except when it occurs at the end of a sentence.

Writing Metric Numbers

- Always use decimals, not fractions (write 0.75 g, not 3/4g).
- Use a zero before the decimal marker for values less than one (write 0.45 g, not .45g).
- Use spaces instead of commas to separate blocks of three digits for any number over four digits (write 45 138 kg or 0.004 46 kg or 4 371 kg). Note that this does not apply to the expression of amounts of money.
- In the United States, the decimal marker is a period; in other countries a comma is usually used.

Conversion and Rounding

- When converting numbers from inch-pounds to metric, round the metric value to the same number of digits as there were in the inch-pound number (11 miles at 1.609 km/mi equals 17.699 km, which rounds to 18 km).
- Convert mixed inch-pound units (feet and inches, pounds and ounces) to the smaller inch-pound unit before converting to metric and rounding (10'-3" = 123"; 123" x 25.4 mm = 3124.2 mm; round to 3124 mm).
- In a "soft" conversion, an inch-pound measurement is mathematically converted to its exact (or nearly exact) metric equivalent. With "hard" conversion, a new rounded, rationalized metric number is used. Products may also be created in new, convenient metric sizes.

Visualizing Metric

A few basic comparisons are worth remembering to help visualize metric measurements:

- One millimeter is about 1/25", or slightly less than the thickness of a dime.
- One meter is the length of a yardstick, plus about 3-1/3".
- One gram is about the mass (weight) of a large paper clip.

- One kilogram is about the mass (weight) of a soft-bound model building code book (2.2 pounds).
- One liter is about the volume of a 4" cube (100 mm x 100 mm x 100 mm).
- One liter of water has a mass of 1 kilogram.
- One inch is just a fraction (1/64") longer than 25 mm (1" = 25.4 mm; 25 mm = 63/64").
- Four inches are about 1/16" longer than 100 mm (4" = 101.6 mm; 100 mm = 3-15/16").
- One foot is about 3/16" longer than 300 mm (12" = 304.8 mm; 300 mm = 11-13/16").
- Four feet are about 3/4" longer than 1200 mm (4' = 1219.2 mm; 1200 mm = 3' 11-1/4").

The metric equivalent of a typical 2' x 4' ceiling grid is 600 x 1200 mm, so metric ceiling tiles and lighting fixtures are about 3/8" smaller in one dimension and 3/4" smaller in the other.

Similarly, the metric equivalent of a 4' x 8' sheet of plywood or drywall is 1200 x 2400 mm, so metric sheets are about 3/4" narrower and 1-1/2" shorter.

"Rounding down" from multiples of 4" to multiples of 100 mm makes dimensions exactly 1.6% smaller and areas about 3.2% smaller. About 3/16" is lost in every linear foot.

Figure I.1 is a table comparing SI metric with the inch-pound system of measure.

Rules and Guidelines for Using Metric

Rules for Linear Measurement (Length)
- Use only the meter and millimeter in building design and construction.
- Use the kilometer for long distances, and the micrometer for precision measurements.
- Avoid use of the centimeter.
- For survey measurement, use the meter and the kilometer.

Rules for Area
- The square meter is preferred.
- Very large areas may be expressed in square kilometers and very small areas, in square millimeters.
- Use the hectare (10 000 m^2) for land and water measurement only.
- Avoid use of the square centimeter.
- Linear dimensions such as 40 x 90 mm may be used; if so, indicate width first and height second.

Rules for Volume and Fluid Capacity
- Cubic meter is preferred for volumes in construction and for large storage tanks.
- Use liter (L) and milliliter (mL) for fluid capacity (liquid volume). One liter is 1/1000 of a cubic meter, or 1000 cubic centimeters.

Preferred Metric Dimensions for Use in Building Construction

In design, production, and construction, the majority of all measurement statements involve *linear measurement*, in the form of requirements for length, width, height, depth, thickness, diameter, circumference, and so on. Frequently, such measurement statements are not independent; they are part of a set or sequence of values. To select the most appropriate metric values during conversion of linear dimensions, it is helpful to appreciate the concept of *dimensional coordination*, which involves special dimensional preferences for buildings and building products. In metric dimensional coordination, a common set of preferred dimensions is used to establish the geometry of buildings as well as the sizes of constituent components or assemblies. Because all preferred dimensions are related to a building module, the term *modular coordination* is sometimes substituted.

In metric building, the fundamental unit of size is the *basic module* of 100 mm. This metric module is designated by the symbol M. It is slightly shorter

Concepts in Measurement: Comparison of SI and U.S. Customary Units		
Concepts	**SI (Metric) Units**	**U.S. Customary Units**
Completeness	SI has a unit for **every** physical quantity.	Multiple units exist for most physical quantities, but are supplemented by metric or SI units in many fields.
Uniqueness	SI has only **one** recognized unit for any one physical quantity. (For practical reasons, a number of non-SI units or multiples are accepted—units for time, angle, and multiples such as hectare, liter, metric ton.)	A variety of units exist for physical quantities, often for special applications only. For example, without counting superseded or special units, there are 9 units for length, 5 for mass, and 7 for volume or capacity.
Coherence	SI units are all coherent; that is, they have a one-to-one relationship to each other, based on their unit derivation according to the laws of physics.	Customary units have only limited coherence, and result in the introduction of factors other than 1, when changing from one unit to another; for example, 2, 3, 4, 6, 8, 9, 12, 16, 20, 22, 24, 27, etc.
Decimalization	SI uses standard **decimal** prefixes (powers of 10) to alter the magnitude of the reference quantity. These prefixes merely change the position of the decimal point, not the digits in a number. The prefixes are internationally used and understood.	Large or small values are transformed into a different, non-decimal unit, except for the mil (1/1000″), the square (100 ft^2), and the kip (1000 lbf). The ratio between units, or factor, alters the numerical values in calculations.
Unit Names	SI unit names, except for traditional metric names taken from Greek or Latin words, are derived from the names of great scientists, which are the same in all languages.	Customary units have names based on words in the English language. These words have no meaning in other languages. In some instances, the same name is used for different magnitudes.
Symbolization and Rules	SI units and prefixes are represented by internationally agreed letter symbols, which have the same meaning regardless of surrounding language or script. SI has agreed rules for the use of units, symbols, and numbers.	Unit names can be represented by symbols (ft, lb, qt), signs (″, ′, :), or abbreviations (fps, psi, cfm). Their use relates to the English language context only. There are few formal rules on unit use.
Reproducibility	SI units are scientifically defined to allow their accurate determination anywhere in the world, except for the kilogram which is based on an artifact.	U.S. customary units have been defined in terms of metric or SI units; e.g., 1 ft = 0.304 8 m.
Universality	SI units are "international" and represent the official measurement system of most nations of the world. An international governing body (CCPM) maintains the system and its rules, including periodic review.	U.S. customary units represent a national measurement system, not used outside the United States.

Figure I.1

than the 4" (101.6 mm) module that has been used in the United States, and should not be equated with this customary module because metric modular product dimensions will be 1.6% shorter.

The basic module of 100 mm has already been endorsed as the basic unit of size in *metric dimensional coordination* in the United States. Preferred dimensions of buildings and preferred sizes of building components should be whole multiples of 100 mm, wherever practicable. The relationship then becomes mutually reinforcing: preferred sizes can be used to the greatest advantage in buildings set out to preferred dimensions, and the design of buildings in preferred dimensions will encourage the procurement and use of building products in metric sizes.

Building products vary from small components placed by hand that range in size up to about 1200 mm, to larger elements placed by mechanical means that may range up to 12 000 mm. Building dimensions vary from small thicknesses of structural elements and dimensions of small spaces to very large spaces with dimensions of 60 000 mm or more in special structures. To ensure efficient use of materials, preferred dimensions play an important part in design, production, and construction.

There are numerical considerations in the determination of preferred dimensions and sizes:

- **Preferred multimodular dimensions**: selected multiples of the basic module
- **Inframodular sizes**: selected dimensions that are smaller than the basic module
- **Intermodular sizes**: dimensions that are larger than 100 mm, but not a whole multiple of the basic module

It is important to appreciate that preferred dimensions in the context of metric dimensional coordination are *reference dimensions* or *ideal dimensions*, rather than actual dimensions. Allowances for joints, tolerances, and deviations are taken into account in the determination of actual dimensions. For example, when a component is described by a preferred size of 400 mm, this dimension includes an allowance for half a joint width on either side of the component, and the *actual dimension* is less to ensure fit in a coordinating space. If the design joint thickness is 10 mm, the dimension for use as *manufacturing target dimension* will be 390 mm.

See Figure I.2 for a detailed chart listing key metric units for use in civil and structural engineering, mechanical engineering, and electrical engineering. Figure I.3 lists the metric units used in the construction trades. In this chart, the term *length* includes all linear measurements (length, width, height, thickness, diameter, and circumference).

Preferred Multimodular Dimensions

It is important to select multimodular preferences carefully. Accepting all multiples of 100 mm as preferences may lead to too many choices, making standardization and variety reduction difficult.

The construction process involves joining many individual and often repetitive components, assemblies, or elements into an organized whole. Therefore, building dimensions that are highly *divisible* multiples of the basic module are superior to prime number multiples.

The first step in selecting preferred multimodular dimensions is to choose composite numbers with the largest number of prime factors. This will allow the widest range of combinations of units to exactly match the preferred dimension. Most small multimodular building components, such as bricks, tiles, blocks, and panels, will have basic sizes of 200 mm, 300 mm, 400 mm, and 600 mm. This means that any multiple of 100 mm, which includes as factors the numbers 2, 3, 4, and/or 6, will be a strong preference because the variety of design options is greatly increased.

Dimensions in which 600 mm, 1200 mm, or 6000 mm are factors will be highly divisible and, therefore, preferred in a dimensionally coordinated building environment. Although dimensions divisible by 500 mm or 1000 mm may appear to yield useful values, dimensions divisible by 600 mm and whole multiples thereof are more useful to the designer in planning and detailing.

The dimension of 6000 mm provides a *supermodule* on which larger building dimensions can be based to ensure maximum factorization.

As a general rule, based on the concept of *divisibility*, all multiples of prime number basic modules over 5 (e.g., 7, 11, 13) should be avoided as controlling dimensions unless they are the only modular options for functional or economic reasons.

Civil and Structural Engineering Conversion Factors			
Quantity	From Inch-Pound Units	To Metric Units	Multiply by
Mass	lb kip (1000 lb)	kg metric ton (1000 kg)	0.453 592 0.453 592
Mass/unit length	plf	kg/m	1.488 16
Mass/unit area	psf	kg/m^2	4.882 43
Mass density	pcf	kg/m^3	16.018 5
Force	lb kip	N kN	4.448 22 4.448 22
Force/unit length	plf klf	N/m kN/m	14.593 9 14.593 9
Pressure, stress, modulus of elasticity	psf ksf psi ksi	Pa kPa kPa MPa	47.880 3 47.880 3 6.894 76 6.894 76
Bending moment, torque, moment of force	ft-lb ft-kip	N·m kN·m	1.355 82 1.355 82
Moment of mass	lb·ft	kg·m	0.138 255
Moment of inertia	lb·ft^2	kg·m^2	0.042 140 1
Second moment of area	in^4	mm^4	416 231
Section modulus	in^3	mm^3	16 387.064

(from Metric Design for Federal Construction, 1991, National Institute of Building Sciences, Washington, DC)

Figure I.2a

Mechanical Engineering Conversion Factors			
Quantity	From Inch-Pound Units	To Metric Units	Multiply by
Mass/area (density)	lb/ft^2	kg/m^2	4.882 428
Temperature	°F	K	5/9(°F − 32) + 273.15
Energy, work, quantity of heat	kWh Btu ft·lbf	MJ J J	3.6 1.055 056 1.355 82
Power	ton (refrig) Btu/s hp (electric) Btu/h	kW kW W W	3.517 1.054 350 745.700 0.293 071
Heat flux	Btu/f^2·h	W/m	3.152 481
Rate of heat flow	Btu/s Btu/h	kW W	1.055 056 0.293 071 1
Thermal conductivity (k value)	Btu/ft·h·°F	W/m·K	1.730 73
Thermal conductance (U value)	Btu/ft^2·h·°F	W/m^2·K	5.678 263
Thermal resistance (R value)	ft^2·h·°F/Btu	m^2·K/W	0.176 110
Heat capacity, entrophy	Btu/°F	kJ/K	1.899 1
Specific heat capacity, specific entrophy	Btu/lb·°F	kJ/kg·K	4.186 8
Specific energy, latent heat	Btu/lb	kJ/kg	2.326
Vapor permeance	perm (23°C)	ng/(Pa·s·m^2)	57.452 5
Vapor permeability	perm/in	ng/(Pa·s·m)	1.459 29
Volume rate of flow	ft^3/s cfm cfm	m^3/s m^3/s L/s	0.028 316 8 0.000 471 947 4 0.471 947 4
Velocity, speed	ft/s	m/s	0.3048
Acceleration	f/s^2	m/s^2	0.3048
Momentum	lb·ft/sec	kg·m/s	0.138 255 0
Angular momentum	lb·ft^2/s	kg·m^2/s	0.042 140 11
Plane angle	degree	rad mrad	0.017 453 3 17.453 3

(from Metric Design for Federal Construction, 1991, National Institute of Building Sciences, Washington, DC)

Figure I.2b

At the international level, and in many national standards, a distinction is made between multimodular dimensions for horizontal applications and for vertical applications. Horizontal dimensions of spaces are generally larger than vertical dimensions and, therefore, benefit from the use of a larger modular increment in the planning module. Dimensional preferences often must be further refined according to individual applications.

Numerical preference in dimensions is illustrated by the following example. When the designer has a free choice of dimensions from 4700 mm to 5000 mm, he or she can list the options concerning whole components of modular

Electrical Engineering Conversion Factors			
Quantity	From Inch-Pound Units	To Metric Units	Multiply by
Power, radiant flux	W	W	1 (same unit)
Radiant intensity	W/sr	W/sr	1 (same units)
Radiance	W/(sr·m^2)	W/(sr·m^2)	1 (same units)
Irradiance	W/m^2	W/m^2	1 (same units)
Frequency	Hz	Hz	1 (same value)
Electric current	A	A	1 (same unit)
Electric charge	A·hr	C	3600
Electric potential	V	V	1 (same unit)
Capacitance	F	F	1 (same unit)
Inductance	H	H	1 (same unit)
Resistance	Ω	Ω	1 (same unit)
Conductance	mho	S	100
Magnetic flux	maxwell	Wb	10^{-8}
Magnetic flux density	gamma	T	10^{-9}
Luminous intensity	cd	cd	1 (same unit)
Luminance	lambert cd/ft^2 footlambert	kcd/m^2 cd/m^2 cd/m^2	3.183 01 10.763 9 3.426 26
Luminous flux	lm	lm	1 (same unit)
Illuminance	footcandle	lx	10.763 9

(from Metric Design for Federal Construction, 1991, National Institute of Building Sciences, Washington, DC)

Figure I.2c

sizes as follows:

1. 4700 mm   2 options   100 mm, 4700 mm
2. 4800 mm   10 options 100 mm, 200 mm, 300 mm, 400 mm, 600 mm, 800 mm, 1200 mm, 1600 mm, 2400 mm, 4800 mm
3. 4900 mm   3 options   100 mm, 700 mm, 4900 mm
4. 5000 mm   6 options   100 mm, 200 mm, 500 mm, 1000 mm, 2500 mm, 5000 mm

	Quantity	**Unit**	**Symbol**
Surveying	length	kilometer, meter	km, m
	area	square kilometer hectare (10 000 m²) square meter	km² ha m²
	plane angle	degree (non-metric) minute (non-metric) second (non-metric)	° ' "
Excavating	length	meter, millimeter	m, mm
	volume	cubic meter	m³
Trucking	distance	kilometer	km
	volume	cubic meter	m³
	mass	metric ton (1000 kg)	t
Paving	length	meter, millimeter	m, mm
	area	square meter	m²
Concrete	length	meter, millimeter	m, mm
	area	square meter	m²
	volume	cubic meter	m³
	temperature	degree Celsius	°C
	water capacity	liter (1000 cm³)	L
	mass (weight)	kilogram, gram	kg, g
	cross-sectional area	square millimeter	mm²
Masonry	length	meter, millimeter	m, mm
	area	square meter	m²
	mortar volume	cubic meter	m³

Metric Units Used in Construction
(from Metric Design for Federal Construction, 1991, National Institute of Building Sciences, Washington, DC)

Figure I.3a

The best alternative is immediately apparent: 4800 mm. This alternative offers 10 possibilities to use whole multimodular or modular components, most of which, in turn, are preferred dimensions.

Based upon these principles, Figure I.4 shows a *set of preferred values* for building geometry and building product sizes. The listings of preferences illustrate the most useful alternatives.

	Quantity	Unit	Symbol
Steel	length	meter, millimeter	m, mm
	mass	metric ton (1000 kg) kilogram, gram	t kg, g
Carpentry	length	meter, millimeter	m, mm
Plastering	length	meter, millimeter	m, mm
	area	square meter	m^2
	water capacity	liter (1000 cm^3)	L
Glazing	length	meter, millimeter	m, mm
	area	square meter	m^2
Painting	length	meter, millimeter	m, mm
	area	square meter	m^2
	capacity	liter (1000 cm^3) milliliter (cm^3)	L mL
Roofing	length	meter, millimeter	m, mm
	area	square meter	m^2
	slope	meter, millimeter	m, mm
Plumbing	length	meter, millimeter	m, mm
	mass	kilogram, gram	kg, g
	capacity	liter (1000 cm^3)	L
	pressure	kilopascal	kPa
Drainage	length	meter, millimeter	m, mm
	area	hectare (10 000 m^2) square meter	ha m^2
	volume	cubic meter	m^3
	slope	millimeter/meter	mm/m

Metric Units Used in Construction (continued)
(from Metric Design for Federal Construction, 1991, National Institute of Building Sciences, Washington, DC)

Figure I.3b

Up to 600 mm, all multiples of 100 mm constitute preferences. In Figure I.5, dimensional preferences above 600 mm are shown in three dimensional ranges:
   a. 600 mm to 3600 mm
   b. 3600 mm to 12 000 mm
   c. Dimensions over 12 000 mm

In the same figure, only three preference categories are shown:
   1. First Preference
   2. Second Preference
   3. Third Preference

	Quantity	**Unit**	**Symbol**
HVAC	length	meter, millimeter	m, mm
	volume	cubic meter	m^3
	capacity	liter (1000 cm^3)	L
	airflow	meter/second	m/s
	volume flow	cubic meter/second liter/second	m^3/s L/s
	temperature	degree Celsius	°C
	force	newton, kilonewton	N, kN
	pressure	kilopascal	kPa
	energy, work	kilojoule, megajoule	kJ, MJ
	rate of heat flow	watt, kilowatt	W, kW
Electrical	length	meter, millimeter	m, mm
	frequency	hertz	Hz
	power	watt, kilowatt	W, kW
	energy	megajoule kilowatt hour	MJ kWh
	electric current	ampere	A
	electric potential	volt, kilovolt	V, kV
	resistance	ohm	Ω

Metric Units Used in Construction (continued)
(from Metric Design for Federal Construction, 1991, National Institute of Building Sciences, Washington, DC)

Figure I.3c

Preferred Multimodular Dimensions from 600 mm to 12 000 mm (in millimeters)		
First Preference	**Second Preference**	**Third Preference**
600	800	700
1200	900	1000
1800	1500	1100 4000
2400	1600	1300 4500
3000	2000	1400 5600
3600	2100	1700 6300
4800	2700	1900 6400
6000	2800	2200 7500
7200	3200	2300 8000
8400	3300	2500 8100
9600	4200	2600 8800
10 800	5400	2900 9900
12 000	6600	3100 10 000
	7800	3400 10 400
	9000	3500 10 500
	10 200	11 200
	11 400	11 700

Figure I.4

Matrix of Criteria for the Allocation of Dimensional Preferences in Building			
Range (mm)	1 First Preference	2 Second Preference	3 Third Preference
A 600 – 3600	All multiples of 600	All multiples of 300 and 400, not included in A.1	All multiples of 100, not included in A.1 and A.2
B 3600 – 12 000	All multiples of 1200	All multiples of 600, not included in B.1	All multiples (except primes) of 300 and 400, not included in B.1/B.2
C Above 12 000	All multiples of 3000	All multiples of 1200, not included in C.1	All multiples of 600 and 1500, not included in C.1 and C.2
All values are in millimeters and represent multiples of the international building module of 100 mm.			

Figure I.5

Section 2
Estimate Considerations

The approach taken in developing this book is that to be successful in estimating in metric, all estimators must learn the metric system, as well as how to estimate as contractors do. If the estimator is thinking about the building process while preparing the estimate, the final estimated price should be accurate. Constructability is the key here. All through the process the estimator should be thinking construction – *How am I going to put this item together? Where will I stockpile the materials?* and so on. After the *How* is answered with a price, it's time to put Overhead and Profit on the estimate. Overhead and Profit are vital to a job being finished. The percentages charged by one contractor undoubtedly differ from those of another contractor because each conducts business differently, has different company markups, and probably does a different volume of work each year.

What Is an Estimate?

An estimate is a reliable cost/time evaluation of an item or project both in "part" and "total," for both "present" and "life cycle." An estimate contains essentially three variables:

- Quantity
- Quality
- Budget

These three variables are tightly woven together and are interdependent.

In *Estimating for Contractors*, Paul J. Cook had this to say about estimating and the estimator: "Estimating is purely mental work, perfectly logical and scientific in form. Its activity occurs between the idea of the designer and the concrete reality of the builder. It interprets the design in a special language called 'cost' which, as one aspect of a construction project, has a kind of reality of its own. Ultimately the exact cost will be made known by the accounting department, but knowledge of it cannot wait; an approximation is necessary before actual physical work begins. The estimator 'reads' the designer's mind and converts the drawing symbols into measurements and man-hours. His are constant choices, made in a consistent manner, toward an imagined whole."

How might we define estimating in relation to the above ideas? Estimating the construction cost of a building is someone's opinion or judgment of the approximate or probable cost.

Who Is Responsible for the Estimate?

A project estimate is the responsibility of at least four different individuals, who are involved in the project at various stages. Each of these individuals has a definite impact on the project estimate.

- Client
- Designer
- Estimator
- Project Coordinator/Construction Manager/Inspector

Common Methods of Estimating

- **Order of Magnitude Estimates**: Based on historical costs per usable unit such as: per hospital bed; per parking space; per apartment. For planning of future projects, this method realizes a + or − 20% accuracy.
- **Square Meter and Cubic Meter Estimates**: Used when only approximate size of the proposed area of the facility is known. Depending on the source of cost information, a + or − 15% accuracy can be expected.
- **Systems Estimates**: Based on both size of the space and other parameters as they pertain to a particular project and the owner's special requirements. Because more specific information is known about the project, a + or − 10% accuracy can be attained from this estimating method.
- **Unit Price Estimates**: From working drawings, specifications, and site visits with the objective of determining the final, detailed estimate. Because detailed information must be available for the estimate, more precision is possible in determining the final estimated cost with a + or − 5% accuracy.

The accuracy mentioned here relates to the average of reasonable bids, assuming the project is competitively bid by contractors. It is not possible, given the variables in the estimating process, to expect precise numbers to come from the estimate. The source of cost information used will have a significant influence on the accuracy of the estimate.

The Quantity Takeoff

Guide for Handling Numbers and Measurements

- Use preprinted forms for orderly sequence of descriptions, dimensions, unit prices, extensions, totals, etc.
- Use only the front side of each piece of paper.
- Be aware of metric dimensioning. Is it based on "hard" metric modules or standard inch-pound products measured in "soft" metric?
- Watch for changes from metric to inch-pound in the same drawing set and in the specifications.
- Be consistent when listing dimensions, e.g., length x width x height.
- Use printed metric dimensions where given.
- When possible, add up printed dimensions for a single entry.
- Measure all other metric dimensions carefully.
- Use each set of dimensions to calculate multiple quantities where possible.
- Do not "round" until the final summary of quantities.
- Mark drawings as quantities are assembled.
- Keep similar items together, different items separate.
- Identify sections and drawing numbers to aid in future checking for completeness.
- Be alert for:
  - Notes such as N.T.S. (Not To Scale).
  - Changes in the scale used throughout the drawings.

- Drawings reduced from the original size.
- Discrepancies between specs and plans.

Quantity Takeoff Procedures

The estimator must have a vision of the sequence of events that result in the completion of a building. Simply stated, it is: Excavate the site, lay a foundation, erect a structure, and cover it with finishes.

The following takeoff procedure is based on that vision:

Concrete

Foundation and Substructure: Broken down into structural excavation and backfill, formwork, reinforcing steel, concrete placement, and finishes. Added work, such as embedded items, joints, water stops, stepped footings, and block-outs or penetrations must also be taken off.

Superstructure: Identify the type of concrete used (e.g., cast in place, job cast tilt-up, plant-produced precast, or job site precast concrete). It is not uncommon for the superstructure to include more than one type. Depending on the type of concrete, the takeoff can be done by the cubic meter, square meter, or linear meter method of measurement. Forming materials and methods vary, and the concrete can be finished using one of numerous methods or left unfinished. Concrete design and color must be given consideration in the takeoff. Remember, concrete strength in the metric system is specified in megapascals (MPa). Structural floor loading capacity is specified in kilograms per square meter (kg/m^2).

When the superstructure is framed with steel, wood, or masonry, cementitious floor and roof fills may be required. Finishes, jointing, block-outs, and closures are part of the takeoff requirements.

Certain building mechanical equipment may require concrete slabs and embedded items that will be necessary to complete the work.

Finally, the equipment required in the placing of forms, reinforcing, concrete, and erecting the precast work must be listed.

Concrete quantities are generally taken off in square meters and millimeters for surface area, cubic meters for volume measurements. Linear measurements are expressed in millimeters or meters.

Masonry

Many different kinds of masonry can be used in any one building, and the method of installation of these materials can vary within the same project. For this reason a detailed takeoff is required to be sure nothing is left out. Pricing masonry work on a square-meter-of-wall basis can be very inaccurate. The masonry should be broken down into exterior walls and interior partitions. Depending on the structural design of masonry units, the reinforcement and grouting requirements vary considerably. Insulation, embedded items, special finishes, and scaffolding and shoring requirements must be identified and listed. Bricks in the metric system will be sized at 190 mm x 90 mm x 57 mm in depth; concrete masonry blocks will be sized at 390 mm x 190 mm x 190 mm in depth. Mortar joints will be 10mm. Area will be expressed in square millimeters or meters.

Structural Steel, Miscellaneous Iron, Ornamental Metals

Structural Steel: A review of the structural drawings will reveal the complexity of the fabrication requirements. The specifications will be the source for the testing and inspection required in the shop or field.

Steel Joists: Identify the type of joist, number of pieces, length, spacing, type of bridging including number of rows and length, number of tie joists, ceiling extensions, top and bottom chord extensions, any special end bearings required, and paint required. Structural steel elements will be taken off in kilograms (kg). The way this is done is to determine the meters of length of the structural steel unit, and multiply this length by its appropriate kilogram weight per meter.

Steel Deck: List the type, depth, gauge, closure, finish, and method of attachment of steel deck. Area will be measured in millimeters or meters.

Composite Studs. List size, length and spacing.

Miscellaneous and Ornamental Metals: Determine the type of finish required. Is the metal furnished and erected, or is it furnished only? Field measurements, shop drawings, the method of installation, and erecting equipment, when required for installation, must all be noted and given consideration. Stairs (type of construction, rails, nosings, number of risers and landings), railings (materials, type of construction, finishes, protection), gratings (types, sizes, finishes), and attachments are included.

Equipment: While making the takeoff per the above procedure, list equipment that will be used to erect fabricated materials.

Carpentry

Carpentry work can be broken down into Rough Carpentry, Laminated Framing and Decking, Finish Carpentry, and Millwork. The rough and laminated materials can be sticks of lumber and sheets of plywood, job-site fabricated and installed. They can also include prefabricated laminated beams and decks, trusses and truss joists, and panelized roof systems, all delivered to the job site and erected.

Finish carpentry materials can be softwood or hardwood sheets of paneling that are job-site cut, fabricated, and installed, or mill-produced cabinets, stairs, railing, trim, and shelving delivered to the job site and erected.

Consideration should be given to the equipment required to erect or install the carpentry materials. Carpentry materials can arrive at the job prefinished or unfinished. Therefore, the man-hours expended to handle, cut, and install the finish materials will vary greatly.

Area is expressed in square meters; linear measurements are in millimeters and meters; and volumetric measurements which are expressed as board feet in the Imperial system, will be quantified in cubic meters.

Moisture Protection

The waterproofing, insulation, sheet metal, roofing, and exterior cladding protect the building from the elements and can be taken off in that order. In the metric system, many of these elements will be taken off in linear meters or square meters.

Doors, Windows, Storefronts, and Finish Hardware

Door and window schedules, floor plans, finish hardware schedules, and building elevations are used for takeoff purposes. Doors, windows, and finish hardware usually are taken off on the same quantity takeoff form, but storefronts and entrance systems are usually subcontracted and taken off separately. This division is relatively easy to take off; it remains a "count" division.

Finishes

Lath and plaster, metal studs, drywall, tile and terrazzo, acoustic ceilings, flooring, painting, and wall coverings are taken off using finish schedules and floor plans. Many items can be grouped together on a single takeoff form, but segregated for cost analysis purposes. This division is taken off using linear meters and square meters.

Specialties, Architectural Equipment, Furnishings, Special Construction, and Conveying Systems

Most of this work is included in the specifications, floor plans, and site improvement plans. It is important to determine how other disciplines may be affected by the special equipment.

Mechanical and Electrical

Mechanical and electrical items are taken off by systems and itemized in linear meter components within the system. For example, plumbing systems have pipe, fittings, valves, and fixtures as components. Pipe lengths and diameters are expressed in millimeters. Temperature is expressed in degrees Celsius. Air distribution systems require hard metric-sized lay-in diffusers and registers. Rectangular metal duct work (available in hard metric) and flexible round duct (soft-converted) are sized in millimeters.

Lighting systems have raceways, conductors, circuit breakers, switches, and lighting fixtures as components. Cable and conduit are quantified in lineal meters. Amps, volts and watts will continue to be used.

Site Work

The site work can be broken down into the following components:

- Site clearing
- Roads and walks
- Earth moving
- Site improvements
- Caissons and piling
- Landscaping
- Site drainage and utilities

In the metric system, linear meters, square meters, cubic meters, and hectares will be the units of takeoff and measurement for many of these elements.

Summary

The preceding is intended to be used as a guide when making quantity takeoffs of the materials used on the project.

When working with the plans, try to approach each job in the same manner. For example:

- Lower floors to higher floors
- Main section, then wings
- Consistently south to north or vice versa
- Consistently clockwise or counterclockwise
- Floor plan first, elevations next, then detail drawings

By keeping a uniform and consistent system, the chance of omission is greatly reduced.

The following shortcuts can be used successfully when making a takeoff:

- Abbreviate where possible.
- List all gross metric dimensions that can be either used again for different quantities or used to rough check other quantities for approximate verifications (exterior perimeter, gross area, individual floor areas, etc.).
- Multiply the large numbers first to reduce rounding errors.
- Consolidate units when the final quantities have been obtained (e.g., in concrete, to m^3).

- Organize the takeoff to achieve maximum future benefits if you are the low bidder. Items to consider are as follows:
  - Scheduling requirements — items per floor or per wing
  - Materials purchases
  - Material ratios for future reference, eg., concrete forms: m^3 of concrete; kg (metric ton) of reinforcing steel: m^3 of concrete; $: m^2 of building area; $: m^3 for various types of concrete, etc.
  - Items should be measured in the same units consistently throughout the construction process — from estimating through field reporting, accounting, and into cost feedback at the end of the job.
- Take advantage of design symmetry or repetition:
  - Repetitive floors
  - Repetitive wings
  - Symmetry about a center line
  - Similar room layouts
- "Outs" should be either written in red or circled so that they will be deducted instead of added.
- When figuring alternates, it is best to total all items that are in the basic bid, then total all items that are included in the alternates, working with positive numbers in all cases. When adds and deducts are used, confusion may result over whether to add or subtract a given item, especially on a complicated or involved alternate.

Extending the Quantity Takeoff

The standard procedure for extending the quantity takeoff depends on developing precision rules consistent with measurement capabilities, per examples. The table below is a sample takeoff for a concrete footing.

DESCRIPTION		NO.	(M) DIMENSIONS L / W / H	Excavation	UNIT	Forms	UNIT	Reinforcing	UNIT	Concrete	UNIT
Ftg. L-1	E	1	24 / 1.2 / 0.6	123	m^3						
	F	2	24 / / 0.3			14.4	m^2CA				
(6m w/30cm lap)	R	3	25 / / 1.55Kg					116	Kg		
	C	1	24 / 0.9 / 0.3							6.5	m^3
Ftg. L-2	E	1	37 / 1.5 / 0.9	50	m^3						
	F	2	37 / / 0.5			37	m^2CA				
	R	3	38 / / 1.55Kg					117	Kg		
	C	1	37 / 1.2 / 0.5							22.2	m^3
				67.3	m^3	51.4	m^2CA	293	Kg	28.7	m^3
				71	m^3			308	Kg	30	m^3

Item	Input	Output
Concrete	nearest mm	nearest m^3
Formwork	nearest mm	nearest m^2CA
Finishing & Precast	nearest mm	nearest m^2
Lumber	nearest mm	nearest m^3
Finishes	nearest mm	nearest m^2
Earthwork	nearest mm	nearest m^3

Use a calculator with at least one cumulative memory to accumulate several different quantities, which will be multiplied (or divided) by some standard dimension. Using concrete footings (formed 2 sides) as an example:
(Note: If end forms are needed, they should be included.)

E: Excavation

F: Forms

R: Reinforcement

C: Concrete Form Ratio = 51.40/30 = 1.71 $m^2 CA/m^3$

This resultant form ratio should be roughly cross-checked with similar jobs as a check against order of magnitude errors.

Pricing the Estimate

Pricing information can come from a variety of sources. If estimating in metric, be sure any published cost data used is also in metric. Sources include:

- Your own current cost records for similar work, modified to the conditions anticipated on the particular job.
- Responsible subcontractor quotations for the job.
- Material quotations from competitive vendors for the job.
- Unit price cost books such as *Means Building Construction Cost Data, Metric Version*, adjusted for local economic conditions.
- Adjusted figures from old projects or unit price cost books. Be sure to convert historical information to metric units.

The various items to be priced should be transferred from the takeoff sheets to pricing sheets, such as the Cost Analysis Sheet shown in Figure I.6 (from *Means Forms for Building Construction Professionals*).

Most companies use the chief estimator or chief cost engineer to price the final estimate. In small companies, the pricing may be done by the president or the owner.

If pricing is done by the estimator, it should be carefully checked for completeness by the chief cost engineer or estimator within the company. Usually a senior company manager will determine the appropriate profit and contingency allowance to add to the estimate. This final adjustment is determined by considering the following items:

- General precision and accuracy of the estimate, which may be a function of how complete and/or complex the plans and specifications are.
- Bidding competition expected.
- Potential risk involved because of the type of construction.
- Workload and bonding capacity of the contractor.
- Any other internal or external factor that could directly or indirectly influence the expected construction cost.

Types of Costs

Material Costs

The material portions of an estimate are generally easier to evaluate than the labor and equipment. In order of reliability, the sources of material costs should be:

1. Vendors' written or telephone quotations. (Be sure to carefully confirm the unit of measure.)
2. Vendors' catalogs that include current metric price information.
3. Company cost records for previous jobs, factored and converted, where necessary, to the metric measurement system.
4. A reliable unit price cost book, such as *Means Building Construction Cost Data, Metric Version*.

Even though the material quantities estimated may be relatively easy to price, several factors must be carefully analyzed to ensure complete and reliable cost information. Items to consider in analyzing material quotations are:

COST ANALYSIS

SHEET NO.

PROJECT | CLASSIFICATION | ESTIMATE NO.

LOCATION | ARCHITECT | DATE

TAKE OFF BY | QUANTITIES BY | PRICES BY | EXTENSIONS BY | CHECKED BY

DESCRIPTION	SOURCE/DIMENSIONS			QUANTITY	UNIT	MATERIAL		LABOR		EQUIPMENT		SUBCONTRACT		TOTAL	
						UNIT COST	TOTAL	UNIT COST	TOTAL	UNIT COST	TOTAL	UNIT COST	TOTAL	UNIT COST	TOTAL

Figure I.6

- How long is the price guaranteed?
- Does the price include a premium for metric production?
- Does the price include delivery?
- Does the price include sales tax?
- Is the material supplied as per plans and specifications?
- Is there an alternate, less expensive material that might be acceptable?
- What is the lead time between ordering and delivery?
- Does the quotation include an escalation clause?
- Have all available discounts been accounted for?
- Are there any unusual payment requirements (cash up-front, etc.)?
- Are the prices being quoted in metric or imperial units?

Labor Costs

For the contractor with reliable and easily accessible cost records, the best starting point is to price the labor from the company cost records or previous jobs adjusted for the conditions expected for the new projects. If historical records in inch-pound units are used to estimate a metric-dimensioned project, care must be taken to ensure that correct metric conversion values are utilized.

If no labor cost records are available, the estimator has three basic options:

- Use a reliable annual unit price cost book with metric converted costs and output rates, such as *Means Building Construction Cost Data, Metric Version*.
- Study "how to" estimating books for similar types of construction and synthesize the crew and daily production for accomplishing the work.
- Synthesize "from scratch" the crew and expected daily output. Be sure to carefully match the units on which the labor and output information is based.

Any historical labor costs, either from the contractor's records or from any of the cost books, should be factored for the expected environment (i.e., metric dimensions) of the new project. This factoring, or adjusting, may be done in one of three ways:

- As each individual unit price is written (most accurate, most time-consuming).
- By division and subdivision totals (fairly accurate, fairly quick).
- By adjusting the total labor figure for the job (least accurate, fastest).

In addition to adjustments for variances from historical information to account for specific project differences, early metric projects may require a small allowance for lost productivity as a result of worker unfamiliarity. The magnitude of productivity loss will be difficult to predict and should be relatively short-lived. Training may be required to minimize the metric learning curve.

Labor costs are the most unpredictable of all costs for a building project. For jobs that run for more than a few months, some allowance for labor rate escalation is normally included.

Equipment Costs

For most building construction projects, the total equipment costs are a small percentage of the total costs as compared with heavy construction (highways, dams, etc.), where the equipment costs are a major percentage of the total costs. For building construction estimating, equipment costs may be figured two ways, the preference being up to the particular contractor.

The two methods of figuring equipment costs are:

- Include the costs directly in each item as a separate and identifiable unit price, totaling up at the end of the estimate after all the unit prices have

been extended. This approach works well for particular items but may not cover major equipment, such as a tower crane hoist, and personnel that may be used for the job's duration.
- Include the cost of all equipment as an overhead item and charge to the total job for the estimated time the equipment will be on the job.

Whatever method the contractor uses, the estimating assumptions and the cost accounting feedback must be consistent so that the estimators can properly identify the historical equipment costs to help in figuring future jobs. Total equipment costs fall into two categories:

- Rental, lease, or ownership costs, which may be figured on an hourly, daily, weekly, monthly, or annual cost.
- Operating costs, which may include fuel, oil, and routine maintenance.

Equipment rental costs may be obtained from the following sources:

- Quotations from equipment suppliers.
- Annual equipment rental rate compilations such as: *Rental Rate Blue Book for Construction Equipment. Rental Compilation for Construction Equipment*, Association of Equipment Distributors.
- Annual construction cost books with full equipment sections, such as *Means Building Construction Cost Data, Metric Version.*

Equipment leasing costs are necessarily long range. The cost figures are available from equipment manufacturers and local dealers handling the particular equipment.

Equipment ownership costs should be available from the company accounting department. The ownership costs should be analyzed in accordance with accepted accounting guidelines as described in various construction equipment manuals.

The ownership cost of equipment must include consideration of the following components:

- Interest rates
- Economic life of the equipment
- Insurance, storage, etc.
- Taxes and licenses
- Maintenance support facilities
- Salvage value at end of economic life

Equipment operating costs apply to leased and owned equipment and, in some cases, to rented equipment, depending on the type of equipment and rental agreement. The operating costs of equipment are available from the following sources:

- The company's own records.
- Annual cost books containing equipment operating costs, such as *Means Building Construction Cost Data, Metric Version* and *Construction Equipment Cost Reference Guide.*
- Manufacturers' estimates.
- Textbooks dealing with equipment operating costs.

The operating costs for rental equipment consist of fuel, lubrication, expendable parts replacement, minor maintenance, transportation, and mobilizing costs.

Subcontractors

Subcontractor quotations should be carefully examined to ensure the following:

- Have correct metric measurement units and subsequent math extensions been correctly applied?
- Are they complete as per the plans and specifications?

- Do they include sales tax?
- Do they involve any unusual scheduling requirements or constraints?
- Do they include a performance bond, or are they bondable?
- Do they include any unusual payment schedule?
- What is the experience of the subcontractor with the exact type of work involved?
- Is the price competitive or is it a "street" price?

Assuming an out-of-town bid location and that the general contractor has made an accurate takeoff, sources of unit price information are:

- The most reliable of the annual cost books, e.g., *Means Building Construction Cost Data, Metric Version.*
- Prices from a "friendly" subcontractor close enough to the location who could conceivably do the work, if required to do so.

After all the unit prices, subcontractors' prices, and allowances have been entered on the cost analysis sheets, the estimate is extended by the cost engineer. In making the extensions, ignore the cents column and round all totals to the nearest dollar. In a column of figures, the cents will average out and be of no consequence. Indeed, for budget estimates the final figures could be rounded to the nearest 10, or even 100, dollars with the loss of only a small degree of precision. Each division is added and the results checked, preferably by someone other than the one doing the extensions.

It is a good point to check the largest items for order of magnitude errors. If the total division costs are divided by the building area, the resultant square meter cost figures can be used to quickly pinpoint areas that are out of line with expected square meter costs.

For soft metric converted projects there should be little or no change in cost for a metric dimensional project.

At this point the estimator has two choices: All further price changes can be made on the cost analysis sheets, or the total prices of each division can be transferred to an estimate summary sheet so that all further price changes until bid time will be done on one sheet of paper.

Consolidated Estimates

The takeoff and pricing methods discussed in the preceding paragraphs described how to:

- Utilize a Quantity Sheet for the material takeoff.
- Transfer the data to a Cost Analysis form for pricing material, labor, subcontractors, and/or equipment items.

Figure I.7 is a sample Consolidated Estimate Form, on which this information can be summarized.

CONSOLIDATED ESTIMATE

SHEET NO.

PROJECT CLASSIFICATION ESTIMATE NO.

LOCATION ARCHITECT DATE

TAKE OFF BY QUANTITIES BY PRICES BY EXTENSIONS BY CHECKED BY

DESCRIPTION	NO.	DIMENSIONS			QUANTITIES				MATERIAL		LABOR		EQUIPMENT		TOTAL	
						UNIT		UNIT	UNIT COST	TOTAL	UNIT COST	TOTAL	UNIT COST	TOTAL	UNIT COST	TOTAL

Figure I.7

Section 3

Using Means Building Construction Cost Data, Metric Version

In response to the executive order mandating that all federal publications and drawings make the transition to metric, R. S. Means produced the first metric version of *Means Building Construction Cost Data* for 1992. This book, which is now updated annually, will grow each year along with the metric movement and will reflect hard conversions of construction materials and equipment as they occur.

This section describes how *Means Building Construction Cost Data, Metric Version* is organized, where to find the information you need, and how to use it most effectively.

Components of the Cost Section

Numbering System
Like the original *Means Building Construction Cost Data* and other Means cost data books, the Metric Version is divided into the 16 major CSI MasterFormat divisions, plus additional divisions for Square Meter and Cubic Meter costs, as well as City Cost Indexes for making location adjustments. These divisions are patterned after the system adopted by the American Institute of Architects, Associated General Contractors of America, Inc., and the Construction Specifications Institute (CSI). This system is widely used by most segments of the building construction industry.

The numbering system is as follows:

- Major subdivisions are shown in reverse type at the top left-hand corner of the title block.
- Major classifications within the subdivision are shown in bold face type next to the three-digit classification number, which is repeated in the right-hand margin.
- Item line numbers are shown by each line item.
- The complete numerical description is the sequence of all three parts above.

Description
The descriptions of items are organized as follows:

- Left indentation is the general description of the item; it may or may not be priced.
- Items indented under the main item refer to it and are usually different sizes of the main item.

- Items are listed more or less alphabetically, with sizes running from small to large or vice versa.

Crew

The *Crew* column indicates the trade or trades and equipment required to install the described item.

- If an individual trade installs the item using only hand tools, the smallest efficient number of tradesmen will be indicated (1 carp, 2 carp, etc.). Listed is the number in the crew followed by a four-letter abbreviation of the trade.
- If more than one trade is required to install the item or if powered equipment is needed to install the item, a crew number will be designated. Listed is a letter followed by either a number or a number and a letter.
- Hand tools such as drills, power saws, and power hammers are carried as a value but not described individually. Larger equipment is identified by size, capability, and daily cost.

A complete listing of crews is carried in the Reference Section of the book and contains the following components:

- Number and type of tradesmen required.
- Number, size, and type of equipment required (if any).
- Hourly labor costs listed two ways: base rate including fringe benefits and billing rate including the installing contractor's overhead and profit.
- Daily equipment costs based on the weekly equipment rental rate, divided by five, plus the hourly operating cost times eight (8) hours. This cost is listed two ways — both bare and with a 10% markup to cover the installing contractor's overhead and profit.
- The total daily labor man-hours for the crew.
- The total bare costs per day for the crew, including equipment.
- The total daily costs of the crew, including the installing contractor's overhead and profit.

Daily Output

The *Daily Output* column indicates the number of units that the designated crew will produce in one eight-hour day. This represents an average figure. Job conditions will determine the actual field productivity. To calculate unit man-hours, divide the Total Daily Crew Man-hours by the Daily Output.

$$\frac{\text{Total Daily Crew Man-hours}}{\text{Daily Output}} = \text{Man-hours/Unit}$$

Man-Hours

The *Man-Hours* column represents the time required to install one unit of work. Unit man-hours are calculated as shown above.

Unit Column

The *Unit* column indicates how the items are typically measured and priced. Abbreviations are often used to indicate these units. If an abbreviation is unfamiliar, look at the "Abbreviations" pages in the back of the book for clarification. All units used in *Means Building Construction Cost Data, Metric Version* are accepted SI metric as endorsed by the Construction Metrication Council.

Bare Costs

The four *Bare Costs* columns show the Material, Labor, Equipment, and Total installed bare costs per unit. They do not include the installing contractor's overhead and profit.

- *Material* costs are average contract purchase prices for the items, including delivery to the job within a 35-km radius around the 30 major cities used to price materials. The estimator must make proper allowances for higher delivery costs outside the 35-km radius. The materials are assumed to be car-lot quantities for projects costing over $500,000. If the project or quantities are smaller, material prices may be slightly higher because of reduced quantity discounts.
- *Labor* costs are derived by dividing the daily crew labor cost by the daily output. Labor costs are based on the average of union wage rates from 30 major U.S. cities.
- *Equipment* costs include not only rental (or ownership) costs, but also operating costs. This figure is the daily equipment cost divided by the daily output.
- *Total* is the arithmetic sum of the material, labor, and equipment columns.

Total Including Overhead & Profit

The final column represents the total price for the item, including the overhead and profit. It is determined in the following manner:

- Material cost is the bare material cost plus 10%.
- Labor cost is calculated by adding overhead and profit (per the billing rate table) to bare labor costs.
- Equipment cost is marked up 10%.
- For purposes of the estimate, often it is best to add an allowance of 10% to the figures, including overhead and profit, for the prime contractor's markup. This would cover the expense of supervising the work of the installing contractor.
- There is no premium included for metric project work.

Labor Rates

The labor rates used in *Means Building Construction Cost Data, Metric Version* are a 30 major city average for union labor. This information appears in the inside back cover of that publication. The estimator may adjust the labor costs by using either the City Cost Index or local, known labor rates. The method is not important, provided the estimator understands what is included in the labor figures used by R.S. Means. The discussion that follows should provide enough guidance for the estimator to calculate labor rates with or without markups.

For each trade there are three labor rates that should be examined and updated as changes occur.

Base Rate

The base rate is the hourly rate on which the actual payroll is figured. Standard deductions such as withholding taxes, FICA, savings plans, hospitalization, etc., are made each week, with the balance going to the employee.

Base Rate Including Fringe Benefits

The base rate including fringe benefits is the sum of the base rate plus all employer-paid fringe benefits such as vacation pay, employer-paid health and welfare, pension, apprentice training, and industry advancement funds. This rate is printed on the inside back page of *Means Building Construction Cost Data,*

Metric Version for building trades, skilled workers, and helpers for 30 major United States cities based on January 1 of the current year.

Billing Rate (Including Installing Contractor's Overhead & Profit)

The base rate including fringe benefits is adjusted to include the contractor's direct and indirect expenses to arrive at a billing rate that may be used for time, material extras, and so forth.

The contractor's direct and indirect expenses are as follows:

- *Workers' Compensation and Employer's Liability.* The rates vary from state to state relative to the safety record in construction in that particular state, based on the previous year's injuries. Rates also vary by trade according to the hazard involved.
- *State and Federal Unemployment Insurance.* The employer's tax is adjusted by a merit rating system according to the number of former employees making application for benefits. A contractor who finds it possible to offer steady work to employees can enjoy a substantial reduction in the unemployment tax rate.
- *Employer-Paid Social Security (FICA).* Current rates should be checked yearly.
- *Builder's Risk and Public Liability Insurance.* The rates charged vary according to the trades involved and the state in which the work is done. Because of the wide range in rates, many contractors carry this insurance on the applicable trade.
- *Installing Contractor's Overhead.* This add-on includes the costs associated with operating a business. These costs do not contribute directly to the physical construction activities of a project, but are nevertheless necessary to stay in business. This overhead should not be confused with the overhead costs of items in the General Conditions portion of construction jobs. These costs are considered project overhead.
- *Installing Contractor's Profit.* This fee, added by the contractor, offers a return on investment plus an allowance covering the risk involved in the type of construction being bid. The profit percentage may vary from 4% on large straightforward projects to as much as 25% on smaller high-risk jobs. Profit percentages are directly affected by economic conditions, the expected number of bidders, and the estimated risk involved in the project. For estimating purposes, *Means Building Construction Cost Data, Metric Version* assumes 10% as being a reasonable, average profit factor.

The sum of these indirect costs can now be calculated and applied to the base rate including fringe benefits as in the example below:

Description	%	Amount	Total
Carpenter Average Hourly Rate			23.35
Workers' Compensation Insurance	19.2	4.48	
U.S. & State Unemployment	7.3	1.70	
Social Security (FICA)	7.65	1.79	
Builder's Risk/Public Liability	1.89	0.44	
Overhead	11.0	2.57	
Profit	10.0	2.34	13.32
Total	57.04	13.32	36.67

This procedure is applied to each trade to determine the appropriate billing rates.

Square Meter and Cubic Meter Costs

Division 17 in *Means Building Construction Cost Data, Metric Version* is included to facilitate rapid preliminary budget estimates.

The cost figures in this division are derived from more than 11,400 projects contained in the Means data bank of construction costs and include the contractor's overhead and profit. Figure I.8 is a sample page from *Means Building Construction Cost Data, Metric Version 1993*. The prices shown do not include architectural or engineering fees or land costs. The figures have been adjusted to January 1 of the current year. New projects are added to the files each year, while projects over ten years old are discarded. For this reason certain costs may not show a uniform annual progression. In no case are all subdivisions of a project listed.

These projects were located throughout the United States and reflect differences in m^2 and m^3 costs resulting from differences in both labor and material costs, plus differences in the owners' requirements. For instance, a bank in a large city would have different features than one in a rural area. This is true of all the different types of buildings analyzed. As a general rule, the projects on the low side do not include equipment or site work. The median figures generally do not include site work either.

None of the figures "go with" any others. All individual cost items were computed and tabulated separately. Thus, the sum of the median figures for plumbing, HVAC, and electrical will not normally total up to the total mechanical and electrical costs arrived at by separate analysis and tabulation of the projects.

Each building was analyzed as to total and component costs and percentages. The figures were arranged in ascending order with the results tabulated as shown. The 1/4 column shows that 25% of the projects had lower costs, 75% had higher. The 3/4 column shows that 75% of the projects had lower costs, 25% had higher. The median column shows that 50% of the projects had lower costs, 50% higher.

There are two occasions when square meter cost estimates are useful. During the conceptual stage when few, if any, details are available, square meter costs make a useful starting point for ballpark budget purposes. Also, after bids are received, the costs can be worked back into their appropriate units for informational purposes. As soon as details become available in the project design, the square meter approach should be discontinued and the project priced according to its particular components. When more precision is required or for estimating the replacement cost of specific buildings, *Means Square Foot Costs* (available only in Imperial units as of December, 1993) should be used.

When using the figures in Division 17, it is recommended that the median column be used for preliminary figures if no additional information is available. When multiplied by the total city construction cost index and then multiplied by the project size modifier, the median figures should present a fairly accurate base figure, which would then have to be adjusted in view of the estimator's experience, local economic conditions, code requirements, and the owner's particular needs. There is no need to factor the percentage figures; these should remain constant from city to city. All tabulations mentioning air conditioning had at least partial air conditioning.

171 | m², m³ and % of Total Costs

		171 000 | m², m³ Costs	UNIT	UNIT COSTS			% OF TOTAL		
				1/4	MEDIAN	3/4	1/4	MEDIAN	3/4
570	2770	Heating, ventilating, air conditioning	m²	44.50	68.50	87	6.30%	7.90%	9.60%
	2900	Electrical		53	78.50	106	7.60%	9.70%	11.60%
	3100	Total: Mechanical & Electrical	↓	126	176	232	17%	21.60%	26.20%
590	0010	**MOTELS**	m²	430	620	800			
	0020	Total project costs	m³	120	170	277			
	2720	Plumbing	m²	42.50	53.50	64.50	9.40%	10.50%	12.50%
	2770	Heating, ventilating, air conditioning		22.50	39	56.50	4.90%	5.60%	8.20%
	2900	Electrical		39.50	50	65	7.10%	8.10%	10.40%
	3100	Total: Mechanical & Electrical	↓	93.50	123	170	18.50%	23.10%	26.10%
	5000								
	9000	Per rental unit, total cost	Unit	19,900	29,900	38,700			
	9500	Total: Mechanical & Electrical	"	4,025	5,700	6,150			
600	0010	**NURSING HOMES**	m²	640	860	1,025			
	0020	Total project costs	m³	173	226	297			
	1800	Equipment	m²	21	27	42.50	2.30%	3.70%	6%
	2720	Plumbing		57.50	70.50	103	9.30%	10.30%	13.30%
	2770	Heating, ventilating, air conditioning		59	82.50	105	9.20%	11.40%	11.80%
	2900	Electrical		65.50	81.50	107	9.70%	11%	13%
	3100	Total: Mechanical & Electrical	↓	153	201	305	22.30%	28.30%	33.20%
	3200								
	9000	Per bed or person, total cost	Bed	24,200	31,100	38,600			
610	0010	**OFFICES Low-Rise (1 to 4 story)**	m²	535	685	900			
	0020	Total project costs	m³	128	180	242			
	0100	Sitework	m²	38.50	67.50	105	5.20%	9.40%	13.70%
	0500	Masonry		18	42.50	80	2.90%	5.70%	8.60%
	1800	Equipment		6.45	11.95	33.50	1.10%	1.50%	4%
	2720	Plumbing		20	30.50	44	3.60%	4.50%	6%
	2770	Heating, ventilating, air conditioning		43.50	61	89	7.20%	10.40%	11.90%
	2900	Electrical		45	62.50	85.50	7.40%	9.50%	11.10%
	3100	Total: Mechanical & Electrical	↓	93.50	139	204	15%	20.90%	26.80%
620	0010	**OFFICES Mid-Rise (5 to 10 story)**	m²	600	730	980			
	0020	Total project costs	m³	135	174	245			
	2720	Plumbing	m²	18.10	27.50	40	2.80%	3.60%	4.50%
	2770	Heating, ventilating, air conditioning		44.50	64	102	7.60%	9.30%	11%
	2900	Electrical		38	54.50	84	6.50%	8%	10%
	3100	Total: Mechanical & Electrical	↓	108	138	230	16.50%	20.50%	25.70%
630	0010	**OFFICES High-Rise (11 to 20 story)**	m²	710	900	1,125			
	0020	Total project costs	m³	149	212	300			
	2900	Electrical	m²	35.50	51.50	79.50	5.80%	7%	10.50%
	3100	Total: Mechanical & Electrical	"	130	166	274	17.20%	21.40%	29.40%
640	0010	**POLICE STATIONS**	m²	890	1,150	1,450			
	0020	Total project costs	m³	215	279	365			
	0500	Masonry	m²	101	145	179	6.80%	10.60%	15.50%
	1800	Equipment		13.65	59	97.50	1.70%	5.20%	9.80%
	2720	Plumbing		50	71	119	5.60%	6.80%	10.70%
	2770	Heating, ventilating, air conditioning		76	101	137	6.40%	10%	11.70%
	2900	Electrical		89	142	179	9.30%	11.70%	14.80%
	3100	Total: Mechanical & Electrical	↓	241	300	395	21.60%	27.50%	33%
650	0010	**POST OFFICES**	m²	710	860	1,125			
	0020	Total project costs	m³	131	171	203			
	2720	Plumbing	m²	31	39.50	49.50	4.20%	5.30%	5.60%
	2770	Heating, ventilating, air conditioning	↓	44.50	59.50	90.50	6.60%	8%	9.80%

Figure I.8

Cost Indexes

All unit price costs (material, labor and equipment) listed in *Means Building Construction Cost Data, Metric Version* represent U.S. national averages and are given in U.S. dollars. The city cost indexes can be used to convert costs to a particular location.

City Cost Indexes

City cost indexes are listed for 162 cities and are related to a 30 major city average of 100. The city cost indexes can be used to adjust costs in *Means Building Construction Cost Data, Metric Version* to the specific city in which the project is located.

Historical Cost Indexes

Figure I.9 lists both the Means City Cost Index based on January 1, 1975 = 100 as well as the computed value of an index based on January 1, 1993 costs. Since the January 1, 1993 value is estimated, space is left to write in the actual index numbers as they become available through the quarterly *Means Construction Cost Indexes* or as published in *Engineering News Record*. To compute the actual index based on January 1, 1993 = 100, divide the Quarterly City Cost Index for a particular year by the actual January 1, 1993 Quarterly City Cost Index. Space has been left to advance the index figures as the year progresses

Adjustments

Using the City Cost Indexes and the Historical Cost Index, the following adjustments can be made.

A. Location Adjustment

To convert known or estimated costs from one city to another, use the City Cost Indexes as follows:

Year		Quarterly City Cost Index 1/1/75 = 100		Current Index Based on 1/1/93 = 100		Year		Quarterly City Cost Index 1/1/75 = 100	Current Index Based on 1/1/93 = 100		Year		Quarterly City Cost Index 1/1/75 = 100	Current Index Based on 1/1/93 = 100	
		Est.	Actual	Est.	Actual			Actual	Est.	Actual			Actual	Est.	Actual
Oct	1993					July	1978	122.4	53.2		July	1960	45.0	19.6	
July	1993						1977	113.3	49.2			1959	44.2	19.2	
April	1993						1976	107.3	46.6			1958	43.0	18.7	
Jan	1993	230.1		100.0	100.0		1975	102.6	44.6			1957	42.2	18.3	
July	1992		227.6	98.9			1974	94.7	41.2			1956	40.4	17.6	
	1991		221.6	96.3			1973	86.3	37.5			1955	38.1	16.6	
	1990		215.9	93.8			1972	79.7	34.6			1954	36.7	15.9	
	1989		210.9	91.7			1971	73.5	31.9			1953	36.2	15.7	
	1988		205.7	89.4			1970	65.8	28.6			1952	35.3	15.3	
	1987		200.7	87.2			1969	61.6	26.8			1951	34.4	15.0	
	1986		192.8	83.8			1968	56.9	24.7			1950	31.4	13.6	
	1985		189.1	82.2			1967	53.9	23.4			1949	30.4	13.2	
	1984		187.6	81.5			1966	51.9	22.6			1948	30.4	13.2	
	1983		183.5	79.7			1965	49.7	21.6			1947	27.6	12.0	
	1982		174.3	75.7			1964	48.6	21.1			1946	23.2	10.1	
	1981		160.2	69.6			1963	47.3	20.6			1945	20.2	8.8	
	1980		144.0	62.6			1962	46.2	20.1			1944	19.3	8.4	
▼	1979		132.3	57.5		▼	1961	45.4	19.7		▼	1943	18.6	8.1	

Figure I.9

$$\text{Known Cost in City A} \times \frac{\text{City B Index}}{\text{City A Index}} = \text{Unknown Cost in City B}$$

For example, if a building cost was $3,500,000 in Boston, Massachusetts, how much would a duplicate building cost in Los Angeles, California?

$$\text{Boston Cost of } \$3,500,000 \times \frac{\text{Los Angeles } 112.7}{\text{Boston } 118.6} = \$3,325,900 \text{ Los Angeles}$$

If the project has a preponderance of materials for any particular division, then the City Cost Index should be adjusted in proportion to the value of the factor for that division.

B. *Time Adjustment with Location Adjustment*

To find the correct cost from a project built previously in a different city, the following formula is used:

Time Adjustment   **Location Adjustment**

$$\text{City B Present Cost} = \frac{\text{Present Cost Index}}{\text{Former Cost Index}} \times \text{Former Cost (City A)} \times \frac{\text{City B Index}}{\text{City A Index}}$$

For example, what would be the estimated cost to build a building in Los Angeles in 1993, if a similar building was built in Boston in 1974 for $3,500,000?

$$\frac{1993 \text{ Index}}{1974 \text{ Index}} \times \frac{\text{Los Angeles City Index}}{\text{Boston City Index}} \times \$3,500,000 = \text{New Cost}$$

$$\frac{230.1}{94.7} \times \frac{112.7}{118.6} \times \$3,500,000 = \$8,081,000$$

(Estimated Los Angeles, 1993)

The Historical Index figures above are compiled from the Means Construction Cost Index Service. This is not a forecast; rather, it serves as a method to estimate current costs based on the market costs of previous years.

Regardless of the cost index used (Means, *Engineering News Record*, etc.), it is recommended that each estimate, as it is performed, should reference the cost index used and its date.

Part II
METRIC ESTIMATING BY MASTERFORMAT DIVISION

liter
kg
mm
Ω
joule
pascal
tonne
Celsius
m³

Part II
Metric Estimating by MasterFormat Division

Part II consists of estimating guidelines for projects in metric, together with information on the status of metrication for various construction materials and products, all arranged according to the 16 divisions of the CSI (Construction Specifications Institute) MasterFormat. Addresses and telephone numbers of product associations and manufacturers can be found in Appendixes B and C.

A sample building project is described in the following section, preceding Division 1. This project will be used to illustrate the process of estimating a project in metric, by division. Each division (excluding Divisions 11 – 14) is followed by a consolidated estimate worksheet that is filled in with the work description and quantities needed to complete the estimate. In some cases, line numbers from *Means Building Construction Cost Data, Metric Version* are provided. Note: These estimate worksheets were prepared using *Lotus 1-2-3®, Version 3.1.*

Sample Project: Repair Garage

The sample building construction project that may be used as a practice exercise for metric estimating is illustrated in Figure II.1.

The project is the construction of a 372 m^2 (4,000 S.F.) repair garage. The client needs a facility to house and maintain its small fleet of vehicles (cars, pick-up trucks, and delivery vans). Currently, the company is using limited leased space about five kilometers from the plant. The garage will be built in a remote corner of a very large, paved parking area.

Site access is excellent. There is plenty of storage and movement area available on the site. The parking area, including the project site, is locked and patrolled by security guards during nonwork hours (nights, weekends, and holidays).

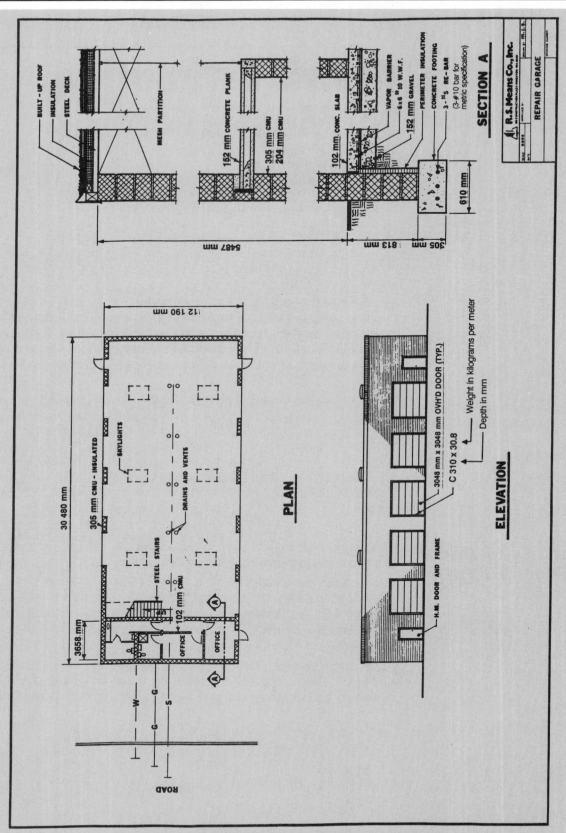

Sample Project Plans (Repair Garage)

Figure II.1

Division 1
General Conditions

Division 1 will not change significantly when metrication is fully implemented. Specifications will be described and detailed in metric units. The remaining sections of the division are basically time durations and dollar values per day, week, or month.

Changes Required for Metric

> Most items covered in this division would not be influenced by the use of metric dimensions for the project. Those that are, such as distances for freight and hauling, areas for scaffolding, temporary protection and cleanup, and permit calculations, should be checked carefully for consistency.

Estimating Procedures

The General Conditions, Division 1, of the specifications provide procedures for performing the contract according to accepted practices in the construction industry.

Although this is the first division in the specifications, it is normally the last to be priced out in the estimating process. Some estimators finish the estimate and then add a percentage to take care of the General Conditions; 10% has become a popular number to use. There are some estimators who price out the General Conditions and arrive at a reasonably precise number. What's the best method? You decide what fits your organization.

This section outlines some of the items included in a contractor's Job Overhead. In performing an estimate for a project, the possible impact of each item should be considered and an allowance included if the item is anticipated on the particular job.

Personnel

Superintendent (Project Manager): Usually priced by the week or month for the job duration.

Field Engineer: Usually priced by the week.

Cost Engineer: Usually priced by the week.

Accountant: Usually priced by the week.

Estimating: Usually priced by the week or month, depending on job requirements.

Warehouse Personnel: Priced on a monthly basis as assigned to the project.

Watchman, Guard Dog, etc.: Some jobs must be watched both day and night, but others require only a security fence, priced by the meter.

Tool Room Keeper: Usually priced by the week or month for some portion of the job duration. Tool maintenance may require an additional allowance on some jobs.

Timekeeper: Usually priced by the week.

Services

Barricades and Signal Lights: Priced as an allowance or estimated quantities if job requirements are complex.

Heat Schedule: Priced by the month. Includes:
- Concrete heat
- Masonry heat
- Finishing heat
- Office and trailer heat

Office Trailers: Rental rates are by the month.

Snow Removal: Usually priced as an allowance based on the expected scope of removal.

Telephone: Usually paid for by the contractor and priced per month based on the job duration.

Temporary Enclosure: Usually priced as an allowance based on estimated quantities of each item from historical cost records.

Temporary Light and Power: Usually priced on the basis of the area of the building and the expected monthly power usage.

Temporary Toilets: Usually priced by the month; the number required is a function of the anticipated size of the work force.

Warehouse or Storage Trailers: Rental rates are by the month.

Contract Provisions

As-Built Drawings: Usually priced as an allowance based on the number of drawings required and the estimated field and office work to produce them.

Final Inspection & Punch List: Normally priced by level of inspection anticipated for the job.

Liquidated Damages: Priced per day of expected time overrun.

Permits: Rates vary from city to city and should be checked with the municipality.

Photographs: Priced by the sequencing required and number of prints, usually as an allowance based on similar, past jobs.

Tree Protection: Normally priced per tree to be protected, or as an allowance.

Signs: Normally priced as an allowance depending on the particular job requirements.

Testing: Normally priced as an allowance based on the size and type of construction.

Labor Provisions

Lost Time Due to Weather: Estimated by climatological data and expected time of the year that the project will be constructed.

Overtime or Premium Time Payments: Usually priced as a percentage of contractor labor and evaluated against the project completion date.

Equipment and General Charges

Broken Glass: Priced as a percentage of the glass and glazing contract.

Cutting and Patching: Usually priced as an allowance based on expected complexity of the project.

Equipment: Cost considerations are buy, lease, or rent, plus operating expense.

Hoisting: May be priced in the individual units or as an overhead item including rental, erection, and operation of the hoists.

Pumping: Normally priced from an analysis of the duration and volume of water expected, which would dictate pump size, power, and labor to install and operate the pumps. This may be an expensive item because pumps may have to operate 168 hours per week during certain phases of construction.

Punch List or Recheck: Usually priced as a percentage of the total project costs.

Rubbish Chutes: Priced per chute. (Rubbish removal from the site is carried as a separate item.)

Temporary Doors: Priced as an allowance.

Temporary Stairways: Usually priced per flight or as an allowance based on the project requirements.

Cleanup

Cleaning Floors and Windows: Priced by the square meter of floors and windows.

Day-to-Day Cleanup: Priced by man-hours required to complete the cleanup.

Rubbish Removal: Priced as to the expected volume. Service contracts are available in most cities.

Temporary Partitions or Screens: Normally priced by the square meter or linear meter.

Office Costs

Prints: Priced as an allowance based on costs of similar jobs.

Drafting: Priced by man-hour.

Engineering: Priced by the number of engineers, technicians, and secretaries for the time assigned to the project.

Job Office Supplies: Usually priced per month of job duration.

Outside Consultants: Priced by the job requirements and proposals from consultants.

The Project Overhead Summary

Project-related indirect costs that are not included in the direct unit costs must be identified and priced out. These items usually find their place in the General Conditions or Project Overhead category of a job since they have no identifiable relationship to the accomplishment of any particular work item. Without overhead items, the job could not be completed. In addition to project-related indirects, some contractors include prorated office expenses and contingency factors in this part of the estimate. On the other hand, many contractors consider main office expenses and contingency factors part of the Overhead and Profit added when the job is totaled.

Division 1 in *Means Building Construction Cost Data, Metric Version* contains data that can be used to complete the Project Overhead Summary when actual contractor cost records are not available to the estimator. The *Project Overhead Summary* shown in Figure II.2 can be used to ensure that items normally included in this division are not omitted.

PROJECT OVERHEAD SUMMARY

PROJECT				SHEET NO.	
LOCATION		ARCHITECT		DATE	
QUANTITIES BY:	PRICES BY:	EXTENSIONS BY:		CHECKED BY:	

DESCRIPTION	QUANTITY	UNIT	MATERIAL/EQUIPMENT		LABOR		TOTAL COST	
			UNIT	TOTAL	UNIT	TOTAL	UNIT	TOTAL
Job Organization: Superintendent								
Project Manager								
Timekeeper & Material Clerk								
Clerical								
Safety, Watchman & First Aid								
Travel Expense: Superintendent								
Project Manager								
Engineering: Layout								
Inspection/Quantities								
Drawings								
CPM Schedule								
Testing: Soil								
Materials								
Structural								
Equipment: Cranes								
Concrete Pump, Conveyor, Etc.								
Elevators, Hoists								
Freight & Hauling								
Loading, Unloading, Erecting, Etc.								
Maintenance								
Pumping								
Scaffolding								
Small Power Equipment/Tools								
Field Offices: Job Office								
Architect/Owner's Office								
Temporary Telephones								
Utilities								
Temporary Toilets								
Storage Areas & Sheds								
Temporary Utilities: Heat								
Light & Power								
Water								
PAGE TOTALS								

Project Overhead Summary Form

Page 1 of 2

Figure II.2a

DESCRIPTION	QUANTITY	UNIT	MATERIAL/EQUIPMENT		LABOR		TOTAL COST	
			UNIT	TOTAL	UNIT	TOTAL	UNIT	TOTAL
Totals Brought Forward								
Winter Protection: Temp. Heat/Protection								
Snow Plowing								
Thawing Materials								
Temporary Roads								
Signs & Barricades: Site Sign								
Temporary Fences								
Temporary Stairs, Ladders & Floors								
Photographs								
Clean Up								
Dumpster								
Final Clean Up								
Punch List								
Permits: Building								
Misc.								
Insurance: Builders Risk								
Owner's Protective Liability								
Umbrella								
Unemployment Ins. & Social Security								
Taxes								
City Sales Tax								
State Sales Tax								
Bonds								
Performance								
Material & Equipment								
Main Office Expense								
Special Items								
TOTALS:								

Project Overhead Summary Form (continued)

Figure II.2b

Division 2
Site Work

There are a number of ways in which estimating for Division 2 will be affected by the implementation of the metric system. Measurement units such as cubic yards, square yards, acres, and other units will of course change to cubic meters, square meters, and hectares. Material hauling and travel distances will be expressed in kilometers when metric is fully implemented. Other factors and more specifics are listed below.

Changes Required for Metric

Metrication of Division 2 tasks requires a new set of dimensions for measuring quantities: square meter units for road surfaces, seeding, and sodding; meters or kilometers for linear dimensions such as guardrail, striping, curb and highway; and cubic meter units for earth, stone, cement, and aggregate.

Signage, that is highway road signs, are also slated for a change in distance expressed, from miles to kilometers. Some state highway departments are specifying dual signs (miles and kilometers); many will replace miles with kilometers. Notice of proposed rule making will be published in the Federal Register in 1993.

Rebar
Bar sizes may change to metric per ASTM A615M, A616M, A617M, and A706M.

AASHTO on Metric
The American Association of State Highway and Transportation Officials (AASHTO) Metric Task Force is developing a *Guide to Metric Conversion* for use by federal and state highway officials. The Task Force found that few rehabilitation activities were affected negatively by metrication. Some tasks, such as taking field measurements, were found to be more efficient in metric.

FHWA on Metric

The Federal Highway Administration (FHWA) approved a metric plan in 1991. September 30, 1996 is their target date for full conversion of the agency's annual $16 billion construction program. FHWA is working closely with AASHTO's Metric Task Force, and is developing its own training course for federal and state highway agency personnel to begin in mid-1993.

Estimating Procedures

Site work tends to be difficult to evaluate. Since every job is different, the estimator must have some feel for the nature of the project. What type of soil – sand, clay, peat, or ledge – must be excavated? On many jobs an assortment of two or three different types of soil may occur. What about boulders? Will any excavated material be reusable as general fill in low areas or as a subbase for paving? These are only a few of the questions that must be addressed in site work estimating.

The site visit is a prerequisite for any thorough estimate. When picking up the bid drawings in a town where the job is to be bid, there is often an opportunity to make visual inspection of the site. Any data supplied on the plans can also be helpful. For instance, borings and their logs provide some technical data on the problems that can be expected.

Preparing the Estimate

The first step in preparing the estimate is to define the work area on the plans. Some site drawings indicate the total area of the parcel, but the work area is defined simply with some shading or by leaving it up to the estimator to calculate. Define this area with a colored pencil. This irregular area needs to be converted into square meters for the purpose of knowing the limits of tree removal, stripping of topsoil, mass grading, building and utility excavation, roadways, parking areas, and landscaping.

For estimating purposes, the fastest and probably the most common method used to convert irregular areas into square meters is called "triangulation," simply dividing the area into a series of triangles and rectangles or squares. The sum or total of their products is the site area.

To Approximate Area:

$$\text{I} = 37.61 \text{ m} \times 40.08 \text{ m} = 1507 \text{ m}^2$$

$$\text{II} = \frac{15 \text{ m}}{2} \times 29 \text{ m} = 218 \text{ m}^2$$

$$\text{III} = \frac{25 \text{ m}}{2} \times 29 \text{ m} = 363 \text{ m}^2$$

$$\textbf{Total} = 2088 \text{ m}^2$$

Site Exploration

Preliminary Soil Investigation, Samplings, and Recommendations

Preliminary sampling is the first form of subsurface site investigation, which will provide the soils engineer with the basic information necessary to prepare a preliminary report. This will include sufficient conclusions and foundation recommendations for the estimator to complete a preliminary cost estimate for planning purposes. It is not intended to be used for the final estimate.

Preliminary investigations can take the form of a hole dug to determine the water table or a steel rod used as a probe to trace out a rock profile. Aside from influencing the extent of the final subsurface testing, preliminary testing can sometimes conclusively eliminate a building site that is unsatisfactory.

Final Soil Investigation, Detailed Exploration, and Recommendations

Detailed explorations are classified into two major groups. First, drive sampling (or dry sampling) is made with a thick-wall sampler and produces a representative but disturbed sample. Second, the undisturbed sample is made with thin-wall tube samples that cause a minimum of physical disturbance.

Drive sampling (dry sampling) is not considered to be an undisturbed sample but is known as a *representative* sample. In practice, such samples are identified and classified in the field and then preserved in moisture-proof containers for further reference or laboratory testing. The sample is taken by actually driving a sampler, or "sample spoon," into the soil at the bottom of the bore hole—hence the term "drive sampling."

Upon completion of the laboratory work, all findings are detailed in the "Final Soils Report," and the conclusions and recommendations included are used for design purposes and preparation of the Architect/Engineer Final Cost Estimate.

During the bidding process of a given project, it is essential that all contractors examine very carefully the "Final Soils Report" and have a complete understanding of the conclusions and recommendations.

Site Clearing and Exploration

Before the excavation process gets under way, land clearing begins. This usually involves the removal of vegetation, including grass, weeds, brush, trees, topsoil, and tree stumps. Some sites require the additional expense of removing boulders, walls and/or buildings with their foundations, or in some cases, abandoned or buried foundations.

Large dozers and special logging machines can save time and should be used whenever a job is large enough to justify their expense. Small stumps can be knocked out of the ground in just minutes; medium-size stumps can be removed in about five minutes; larger, oversize stumps require an hour or more. A good clearing team consists of a heavy tractor dozer assisted by a smaller shovel dozer. Another good stumper is the hydraulic excavator (backhoe).

After the trees have been removed, the topsoil is stripped from all areas that will undergo site work. The topsoil is usually stockpiled on site for later use.

Clear and Grub

The first phase of preparing a site clearing estimate is to visit the site and check the area for accessibility, topography, and wetness or dryness. The next step is to make a tree count of a representative area of the site. Make at least three counts at random for each vegetation type. To conduct these counts, randomly locate two points approximately 100 m apart. Count and measure the growth along a straight line between these points 5 m on both sides. This gives the population of 1000 m^2 + or −.

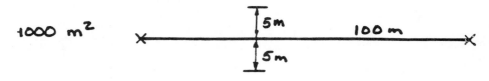

49

Grading

Topographical plans have many uses, but their chief purpose for construction is to provide site data in its most convenient form for planning building projects. For economy, construction projects must be planned to conform closely to existing topography; in other words, to require the least amount of work.

In site development the cost of earthwork is the item most affected by the relationship between the proposed plan and the existing topography. Before a plan is accepted, cost estimates of earthwork for several proposed locations are made in order to arrive at the best and most economical solution.

Swell and Shrinkage

The density of earth undergoes considerable change when the earth is excavated, hauled, placed, and compacted. Because of these changes, it is necessary to specify whether the volume is measured in its original position, in loose condition, or in the fill after compaction.

- Bank Measure is the volume of earth measured in the borrow pit prior to loosening. Payment is usually based on this measure.
- Loose Measure Volume is the volume of the earth after it has been deposited in trucks.
- Compacted Volume is the volume of earth after it has been placed in a fill and compacted.

The volume should be expressed in cubic meters regardless of whether it is bank measure, loose, or compacted.

The weights for earth are usually expressed in kg per cubic meter per the following formula:

Swell: $Sw = \left(\frac{B}{L} - 1\right) \times 100$

Shrinkage: $Sh = \left(1 - \frac{B}{C}\right) \times 100$

Sw = % Swell $\quad$ B = Weight of undisturbed bank
Sh = % Shrinkage $\quad$ L = Weight of loose earth
$\quad\quad\quad\quad\quad\quad\quad\quad$ C = Weight of compacted earth

Example: Find the % swell and % shrinkage for earth whose weights are:

Bank measure undisturbed $\quad$ 1474 kg/m^3
Loose measure $\quad$ 1218 kg/m^3
Compacted $\quad$ 1730 kg/m^3

Swell: $Sw = \left(\frac{B}{L} - 1\right) \times 100 = \left(\frac{1474}{1218} - 1\right) \times 100 = 21\%$

Shrinkage: $Sh = \left(1 - \frac{B}{C}\right) \times 100 = \left(1 - \frac{1474}{1730}\right) \times 100 = 15\%$

Equipment Size

For estimating purposes it is essential that the earthwork equipment selected by the estimator be of a size that matches the excavation operation to achieve maximum production at the minimum cost. The quantity of material to be excavated, the type of material, the haul distance, and the time allowed to complete the job are major factors that the estimator must consider when determining the size of equipment used in the estimate.

Figure II.3 is the takeoff for the site work portion of the sample project.

COST ANALYSIS

PROJECT: Repair Garage
LOCATION:
TAKE OFF BY: ABC
CLASSIFICATION: Division 2
ARCHITECT:
QUANTITIES BY: ABC
PRICES BY: As Shown
EXTENSIONS BY: DEF
CHECKED BY: GHI
SHEET NO. 1 of 15
ESTIMATE NO:
DATE: 1993

DESCRIPTION	SOURCE			QUANT	UNIT	MATERIAL COST	MATERIAL TOTAL	LABOR COST	LABOR TOTAL	EQUIPMENT COST	EQUIPMENT TOTAL	SUBCONTRACT COST	SUBCONTRACT TOTAL	TOTAL COST	TOTAL TOTAL
Site Work															
Saw Cut Asphalt	020	728	0010	220	m	0.72	158	1.43	315	1.11	244				
Remove Asphalt	020	554	1710	2678	m2			1.45	3883	2.05	5490				Backhoe
Haul Asphalt - 3.2 km	022	266	0400	353	m3			0.97	342	2.41	851				
Bulk Excavation	022	238	1200	107	m3			0.62	66	0.84	90				F. E. Loader
Utility Excavation	022	258	0700	111	m			1.73	192	0.74	82				
Footing Excavation	022	254	0050	96	m3			2.97	285	1.74	167				Backhoe
Backfill Utility	022	258	1700	111	m			2.43	270	0.96	107				
Backfill Foundation	022	204	1300	96	m3			0.29	28	0.91	87				
Compact, Vib. Plate	022	204	0600	96	m3			3.23	310	1.28	123				
Borrow Gravel	022	212	0100	57	m3	4.58	261	1.26	72	3.50	200				Dozer
Handgrade Finish	025	122	1150	373	m2			0.79	295	0.08	30				
Concrete Apron, 150 mm	025	120	0020	2344	m2							19.70	46177		Grader/Paver
Subtotal							419		6058		7470		46177		

Note: m2 = square meter   m3 = cubic meter

Figure II.3a

COST ANALYSIS

PROJECT:	Repair Garage	CLASSIFICATION:	Division 2 (cont'd)			SHEET NO. 2 of 15
LOCATION:		ARCHITECT:				ESTIMATE NO:
TAKE OFF BY: ABC	QUANTITIES BY: ABC	PRICES BY: DEF	EXTENSIONS BY: DEF		CHECKED BY: GHI	DATE: 1993

DESCRIPTION	SOURCE		QUANT	UNIT	MATERIAL		LABOR		EQUIPMENT		SUBCONTRACT		TOTAL	
					COST	TOTAL	COST	TOTAL	COST	TOTAL	COST	TOTAL	COST	TOTAL
Site Utilities														
Sewer Line, 150 mm PVC	027	168 2040	49	m	5.40	265	Material + 10%	-------->		370				
Water Service, 38 mm	026	686 2650	43	m	1.64	71	B - 20 Crew 1 Day	-------->		794				
PVC Cl.160, SDR 26														
Gas Service, 51 mm	By Utility Co.										LS	1500		
Electric Service	"										LS	1500		
(Ovhd. to bldg.)														
Telephone	"										LS	500		
Subtotal														4664

Figure II.3b

COST ANALYSIS

PROJECT:	Repair Garage		CLASSIFICATION:	Division 2 (cont'd)					SHEET NO.	3 of 15
LOCATION:			ARCHITECT:						ESTIMATE NO:	
TAKE OFF BY:	ABC	QUANTITIES BY: ABC	PRICES BY: As Shown		EXTENSIONS BY: DEF				DATE: 1993	
									CHECKED BY: GHI	

DESCRIPTION	SOURCE			QUANT	UNIT	MATERIAL		LABOR		EQUIPMENT		SUBCONTRACT		TOTAL TOTAL
						COST	TOTAL	COST	TOTAL	COST	TOTAL	COST	TOTAL	COST
Equipment Mobilization														
Backhoe	022	274	0900	1	Ea.	—		43.00	43	205.00	205			
Front End Loader	022	274	1200	1	Ea.	—		34.50	35	164.00	164			
Dozer	022	274	0020	1	Ea.	—		39.00	39	184.00	184			
Grader (similar)	022	274	1300	1	Ea.							263.00	263	
Paver (similar)	022	274	0700	1	Ea.							350.00	350	
Subtotal							419		117		553		613	
Page 1 Subtotal									6058		7470		46177	
Page 2 Subtotal													4664	
Page 3 Subtotal											553		613	
Division 2 Total							$419		$6,174		$8,023		$51,454	

Figure II.3c

Division 3
Concrete

Concrete costs, like all construction division costs, are composed of material, labor, and equipment. When estimating, units pertaining to each cost should be systematically compiled. Concrete estimating need not be complicated, but it must be detailed. Once an efficient, consistent estimating system is developed, it becomes the basis for job cost records to be used on future estimates.

Quantity takeoff and estimates for Division 3 will be affected by metrication in a number of ways. Ready-mixed concrete will be delivered by the cubic meter (m^3), which is approximately 1.3 cubic yards. Concrete strength grades expressed in megapascals will result in a regular progression, with fewer increments than at present.

Changes Required for Metric

Concrete
The strength designation for concrete will change from psi to megapascals, rounded to the nearest 5 MPa. Actual strength requirements will stay the same.

Rebar
When the metric system is fully implemented, fewer sizes of reinforcing bars will be used. Eleven bar sizes are currently used; eight sizes will be used in the metric system.

Bar sizes may be modified, based on ASTM A615M, A616M, A617M, and A706M. Conventional bars (customarily designated #3 to #18) will be replaced by metric equivalents. The 100 mm^2 rebar will replace both #3 and #4; the 700 mm^2 rebar will replace both #9 and #10. Only #7 (7/8") has no metric equivalent; its use should be limited as the transition period approaches to avoid the possibility of nonavailability after the change is implemented.

ACI on Metric
The American Concrete Institute (ACI) has prepared Building Code 318 according to metric standards. A plan is in place to publish all of ACI's documents in hard metric by 1998.

Estimating Procedures

Concrete estimating can be divided into three distinct groups:

Type I: Cast-in-Place Formed Concrete
- Footings (Continuous and Spread)
- Piers
- Walls
- Columns
- Beams, Elevated
- Underpinning

Type II: Cast-in-Place Slabs and Other Finished Surfaces
- Slab on Grade
- Elevated Slab
- Stairs
- Curbs and Gutters

Type III: Precast Concrete and Cementitious Decks
- Job Site Precast
- Plant-Produced Precast

Skills vary so much that subcontractors usually specialize in one of these trades. Our unit price estimating analysis will approach each of these specialties from the point of view of the installing subcontractor.

Type I: Cast-in-Place Formed Concrete

Although the concrete and reinforcing material costs for a grade beam on the ground and for an exterior spandrel beam on the 20th floor may be the same, forming, placing, stripping, and rubbing costs are not.

Spread Footings

A building designed with spread footings will generally have a footing schedule on the plans defining and dimensioning each type of footing used. For example:

Footing Schedule

Type	No.	Size	Reinforcing
F1 (interior)	8	2438 mm x 2438 mm x 305 mm	9-#6 e.w.
F2 (exterior)	12	1829 mm x 1829 mm x 406 mm	8-#5 e.w.
F3 (corner)	4	1372 mm x 1372 mm x 305 mm	5-#5 e.w.

When the schedule is provided, it eliminates the repetition of detailing on the plans and simplifies the work of the estimator. Transfer the dimensions directly onto the Quantity Sheet. Using this form and the Footing Schedule, the procedure is as follows:

1. Write down the number of each size and the dimensions of each of the spread footings shown on the plans.
2. Extend each of these lines to obtain the volume, form area, and finish area.
   Volume = L x W x D
   Form Area = (2L + 2W) x D
   Finish Area = L x W
3. Total each column.
4. Count the total number of footings on the plan to make sure none have been left out.
5. As you are working on these plans for spread footings, pick off any other items such as anchor bolts, templates, base plates, and reinforcing steel. Reinforcing steel can most conveniently be transferred to a separate sheet. These other items should be computed and totaled on the same Quantity Sheet from which they were derived to avoid errors and omissions.

Continuous Wall Footings and Grade Beams

Wall footings are found under concrete and masonry walls. Grade beams are stiff self-supporting structural members that support wall loads and carry them across unacceptable soil to column footings, caissons, or support piles. The procedure for quantity takeoff is the same as for spread footings.

To accommodate changes in finished grade elevations, wall footings frequently are stepped. These steps will require an additional volume of concrete as well as forming. See Figure II.4 for typical stepped continuous footing.

Proceed as follows:

1. Write down the location and dimensions of each wall footing or grade beam shown on the plan.
2. Extend each of these lines as in the procedure for spread footings to obtain the volume, form area, and finish area.
    Volume = L x W x D
    Form Area = 2L x D
    Finish Area = L x W
3. Total each column.
4. Check off each wall footing and grade beam on the plans to ensure that none have been omitted.
5. Because you have the location and dimensions of these footings at hand, now is the time to take off other items shown that incorporated these dimensions; e.g., reinforcing steel, keyways, anchor bolts, and

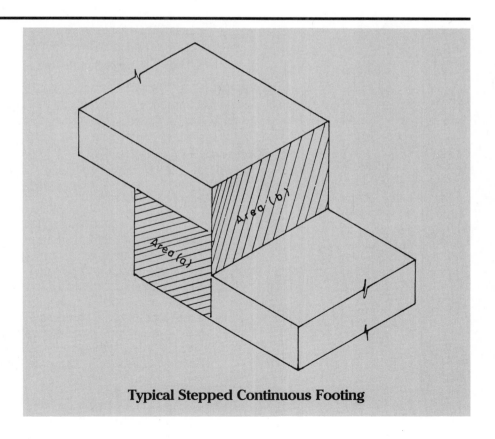

Typical Stepped Continuous Footing

Figure II.4

underdrain. Again, these should be computed and totaled on the same Quantity Sheet from which the dimensions were obtained to avoid errors or omissions.

No discussion of concrete footings would be complete without mentioning "pouring wild." The concrete is placed directly into the excavated trench without forming, thus saving forming material and labor. Check with your local building inspector before attempting such a construction technique. The quantity takeoff procedure is the same as above except that an additional 5% to 10% extra volume of concrete should be allowed for waste.

Pile Caps

Pile caps are basically a spread footing on piles. Their shape is not always rectangular so that the form area is determined by the perimeter times the depth. No concrete volume should be deducted for the protrusion of the piles into the cap.

Piers

Piers are used to extend the bearing of the superstructure down to the footing. They vary in length to accommodate subsurface conditions, original site grades, and frost penetration.

For rectangular and square piers, write down the number of each size and the dimensions of each pier. Extend each line to obtain the volume and form area.

$$\text{Volume} = L \times W \times H$$
$$\text{Form Area} = (2L + 2W) \times H$$

1. Total each column.
2. Count the total number of piers to make sure none have been left out.
3. This is the time to take off the reinforcing steel, anchor bolts, base plates, and templates. Enumerate these on the same Quantity Sheet.
4. For round piers, follow the procedure for rectangular piers exactly except that volume now will be computed from ($\pi R^2 H$).

Walls

Walls are broken down in heights during takeoff. Generally, this is in 1200 mm increments such as under 1200 mm; 1200 mm to 2400 mm; or 2400 mm to 3650 mm. This is because there are different forming and placing costs connected with each of these heights. For large high walls with repetitive selections, consider the use of Gang Forms as described in 031-182, Lines 5500 & 5550, in *Means Building Construction Cost Data, Metric Version*.

1. Using a Quantity Sheet, define each wall. Note that if all walls have footings, the wall footing schedule can be used for this.
2. Write in the dimensions for each of these walls.
3. From these dimensions, compute the volume and form area for each wall, dividing them into convenient modular categories. Note that the brick shelf and slab seat are listed here and have a negative concrete volume.
4. Total each of these columns.
5. While you're looking at the wall sections and details, list everything you see. Now is the time to list beam seats, anchor bolts, base plates, chamfers, pilasters, rustications, architectural finishes, and reinforcing steel. When figuring pilasters, figure the wall forming straight through and then add the area of the pilasters. The additional materials will be used in waste, bracing, and framing.

The above procedure will be identical for Foundation Walls and Retaining Walls. If retaining walls are battered, be certain to note this in identifying them as the erection time is some 20% slower. For special architectural effects, the use of form liners, rustication strips, retarders, and sandblasting must be considered.

See Figure II.5 for a foundation detail.

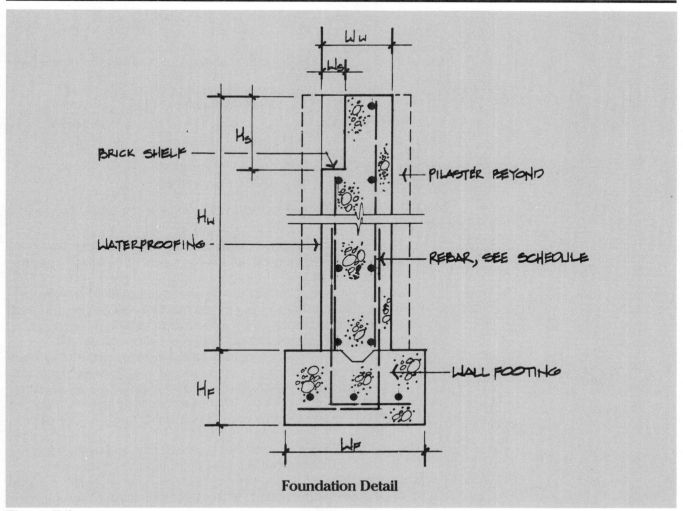

Figure II.5

Columns

Columns, like spread footings, are usually listed in a schedule.

Column Schedule

Identification	No.	Size	Reinforcing
C-1	8	610 mm x 610 mm x 3.7 m	8-#11 ties, #4 @ 610 mm
C-2	12	508 mm x 508 mm x 3.7 m	8-# 9 ties, #3 @ 458 mm
C-3	4	407 mm x 407 mm x 3.7 m	8-# 9 ties, #3 @ 407 mm

Using the Quantity Sheet, follow exactly the same procedure as that described under rectangular, square, and round piers. If column capitals are used, list the number and size required. If columns are to be exposed, be certain to include the finishing, which will be the same as the contact area. Special items to remember about columns are:

1. Measure from floor slab below to underside of beam or slab.
2. Consider multiple use of forms (*Means Building Construction Cost Data, Metric Version*, Forms in Place).
3. Chamfer strips.
4. Finishing Capitals.
5. Reinforcing Steel.
6. Clamps, Elephant Trunk, Inserts, and Placing Pockets.
   Volume = No. x L x W x D
   Form Area = No. x (2 x L + 2 x W) x D

Beams, Elevated

Beam schedules, like footing and column schedules, are usually shown on plans.

Beam Schedule

Identification	No.	Size	Reinforcing
B-1 Ext.	2	305 mm x 762 mm x 23 m	See Rebar Schedule
B-2 Ext.	2	305 mm x 711 mm x 31 m	"
B-3 Int.	4	203 mm x 457 mm x 23 m	"
B-4 Int.	2	152 mm x 457 mm x 31 m	"

Below are some basic principles that must be followed for efficiently estimating C.I.P. beams.

1. Beam depths given in beam schedules are design depths. Deduct the slab thickness when computing contact area and volume for interior beams.
2. List exterior and interior beams separately. Figure II.6 shows that an exterior beam needs forming for the full depth (h) on one side. Also, the access and bracing of this exterior form make it more costly.
3. Figure forming straight through columns as this material will be wasted. Remember columns were figured to the underside of beams so both forming and volume computations go straight through columns.
4. C.I.P. beams and slabs are often placed simultaneously. By keeping the concrete volume for the beams on each floor separate, the combined beam and slab placement cost can be figured for each floor.
5. If you prefer, keep quantities for beam bottoms and beam sides separate. Job costs, however, are difficult to separate this way.
6. Decide upon the method of form support—hung or shored. If shored, remember to allow enough shores for reshoring after stripping.
7. Remember finishing and chamfers for exposed beams.

8. Multiple use of forms depends upon design and finish required.
9. Using a quantity takeoff sheet, write down the identification number and dimension of each type beam.
10. For exterior beams write down both h_e and h_e' dimensions, and for interior beams, h_i' only (Figure II.6). Remember that the beam depth (h_e) in the schedule includes the slab thickness (t).
11. Now extend the computations on each line to obtain the form area and volume:
    Form Area = No. x $(W_e + h_e + h_e')$ x L Exterior
    　　　　　　No. x $(W_i + 2h_i')$ x L Interior
    Volume　　= No. x W_e x h_e' x L + No. x W_i x h_i' x L
12. Total each column on the form in preparation for pricing.
13. Check the total number of beams on the plans to ascertain that none have been left out.

Now is the time, while you are familiar with the beam details, to take off the reinforcing steel or confirm it on the bar list and note any inserts such as hangers.

Underpinning

Although underpinning is done with formed concrete, much of the cost involved is for excavation and shoring. Because of the unique features of each individual job, it's best to analyze the method, labor, and material you propose to use. Division 021-564 of *Means Building Construction Cost Data, Metric Version* provides some good ranges for checking your figures.

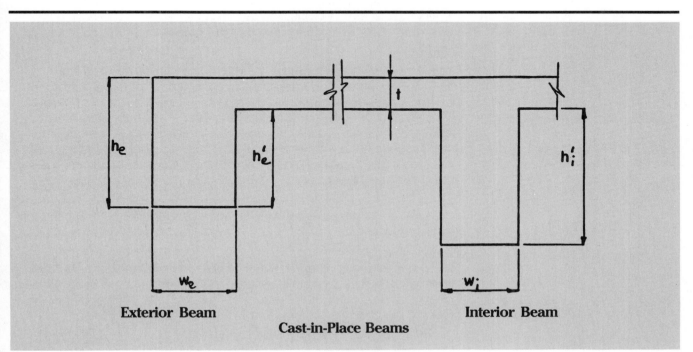

Figure II.6

Forming Type I: Cast-in-Place Formed Concrete

- Footing forms are usually figured with 51 mm stock and 51 mm x 102 mm stakes for minimum waste and maximum reuse. If you intend to use this system, allow 6.6×10^{-2} m^3 per m^{2}CA. Allow 10% waste for each reuse.
- Piers below grade can have many form reuses because they will be buried. The most common method for square tied piers is to use plyform and column clamps. Allow 1.2 m^2 plyform per m^{2}CA plus 3.17×10^{-2} m^3 for bracing lumber per m^{2}CA. Allow 10% waste for each reuse.
- Wall forms come in all sizes, shapes, and materials. Decide which system or systems you're going to use; specify this in your future cost records after you are awarded the contract.
- Columns are formed just like piers except that the reuse of forms will depend upon the finish desired. Good records of the systems, time, and materials used are invaluable in pricing these items.
- Beam and girder forms can be divided into two categories: hung from encased structural steel and shored. The material is the same: 1.15 m^2 of plyform per m^{2}CA and 5×10^{-2} m^3 framing per m^{2}CA; but the fabricating erection and accessories are different. Determine the number of reuses anticipated and allow 10% waste for each reuse. For shored beams and girders, allow enough shores for reshoring after stripping.
- Forming for underpinning is usually one-sided with the earth forming the other side. Allow 1.25 m^2 of plyform per m^{2}CA and at least 1.3×10^{-1} m^3 per m^{2}CA for framing and bracing.
- Slipforms are a special application to which we cannot assign specific quantities per m^{2}CA. For cost ranges, see *Means Building Construction Cost Data, Metric Version*, Slipforms.

Pricing Type I: Cast-in-Place Formed Concrete

There is no better pricing information than your own job cost records adjusted for labor, equipment, and material costs. However, if you plan to use your own records, be sure that all units of measurement agree. If your cost records are incomplete, certain production factors contained in *Means Building Construction Cost Data, Metric Version* will be helpful.

Reinforcing

Reinforcing steel prices for these same columns can be obtained from 032-107-0200 and 0250. The installation cost includes that of the accessories (032-102), but not their material costs. Material and labor prices for splices (032-109) must be added where specified. Material prices for reinforcing steel are very volatile and a local quote should always be obtained prior to using the prices for a bid. For best results, check your rates against the labor rates used in the installation crew. The current 30 city average rate for rodmen is shown inside the back cover of *Means Building Construction Cost Data, Metric Version*.

Charts listing size designations, weights, and diameters for reinforcing bars are shown in Figure II.7.

Placing

The cost of placing concrete is discussed and an analysis of the added cost per m^3 for wheeled concrete is included in R033-090; Placing Ready Mixed Concrete in *Means Building Construction Cost Data, Metric Version*. If we have separated our concrete computations in our initial analysis for volume, m^{2}CA, and finish area, we can now assign placing costs from 033-172 to each volume in our estimate.

Bar Size Designation	Weight Pounds per Foot	Nominal Dimensions — Round Sections		
		Diameter Inches	Cross-Sectional Area-Sq. Inches	Perimeter Inches
#3	0.376	0.375	0.11	1.178
#4	0.668	0.500	0.20	1.571
#5	1.043	0.625	0.31	1.963
#6	1.502	0.750	0.44	2.356
#7	2.044	0.875	0.60	2.749
#8	2.670	1.000	0.79	3.142
#9	3.400	1.128	1.00	3.544
#10	4.303	1.270	1.27	3.990
#11	5.313	1.410	1.56	4.430
#14	7.650	1.693	2.25	5.320
#18	13.600	2.257	4.00	7.090

Bar Size Designation	Weight Kg per Meter	Nominal Dimensions - Round Sections		
		Diameter cm	Cross-Sectional Area - cm^2	Perimeter cm
#3	0.558	0.953	0.71	3
#4	0.992	1.270	1.27	4
#5	1.549	1.587	1.98	5
#6	2.231	1.905	2.85	6
#7	3.036	2.222	3.88	7
#8	3.966	2.540	5.06	8
#9	5.049	2.865	6.44	9
#10	6.391	3.226	8.16	10
#11	7.891	3.581	10.07	11
#14	11.362	4.300	14.51	14
#18	20.199	5.732	25.79	18

R032-020 Metric Rebar Specification - ASTM A615M

Grade 300 (300 MPa* = 43,560 psi; +8.7% vs. Grade 40)				
Grade 400 (400 MPa* = 58,000 psi; -3.4% vs. Grade 60)				
Bar No.	Diameter mm	Area mm^2	Equivalent $in.^2$	Comparison with U.S. Customary Bars
10	11.3	100	0.16	Between #3 & #4
15	16.0	200	0.31	#5 (0.31 $in.^2$)
20	19.5	300	0.47	#6 (0.44 $in.^2$)
25	25.2	500	0.78	#8 (0.79 $in.^2$)
30	29.9	700	1.09	#9 (1.00 $in.^2$)
35	35.7	1000	1.55	#11 (1.56 $in.^2$)
45	43.7	1500	2.33	#14 (2.25 $in.^2$)
55	56.4	2500	3.88	#18 (4.00 $in.^2$)

* MPa = megapascals

Imperial and Metric Sizes and Weights of Reinforcing Bars

Figure II.7

Material

Material prices are developed from R033-060; Concrete Material Net Prices, R033-070; Ready Mix Material Prices, and R033-080; Field Mix Concrete in *Means Building Construction Cost Data, Metric Version.* Local and national averages are listed but these should always be checked against current local prices. Choose the strength required and arrive at a price from 033-122 and -126.

Type II: Cast-in-Place Slabs and Other Finished Surfaces

Several items should be considered when estimating a Slab on Grade.

- Granular Base
- Fine Grade
- Vapor Barrier
- Edge Forms
- Expansion Joints
- Contraction Joints
- Welded Wire Fabric and Reinforcing
- Screeds
- Concrete Material
- Finish and Topping
- Cure and Harden
- Concrete "Outs"
- Haunches

Although this looks like a long and ominous list, all these quantities are derived from a few dimensions. The principal object is not to leave any of these items out of your estimate by oversight. All of the items above may not be incorporated in each Slab on Grade to be estimated, but the checklist is a good reference.

Some basic rules to remember when estimating a Slab on Grade are:

- Allow about 25% compaction for granular base.
- Allow 10% overlap for vapor barrier and welded wire fabric.
- Allow 5% extra concrete for slab on grade.
- No deductions for columns or "outs" under 1 m^2.
- If screeds are separated from forming and placing costs, allow 300 mm of screed per 1 m^2 of finish area.

Building Slab on Grade

Enter on the Quantity Sheet each of the quantities related to Slab on Grade. All the quantities, except for expansion joint around columns, are derived from the three basic dimensions L, W, and t of the slab being estimated. When filling in quantities on the Quantity Sheet, fill in each line and draw a line through the space when there is no quantity. This helps eliminate omissions and duplications. Be sure to add the concrete volume, hand excavation, and reinforcing steel where haunches are shown on the plan.

Sidewalk and Paving

More edge forms and joints will be in sidewalk and paving concrete than in the slab on grade in the building. Other than this and the type of finish normally provided, the estimating procedure is the same.

Elevated Slab

In estimating elevated slabs, we use the same basic dimensions—L, W, and t—as we used in estimating our slab on grade. The slab form area is usually figured straight across the beams, which makes it the same as the finish area. Pay close attention to edge forming around openings such as stairwells and elevator shafts. Also, these can contribute a significant "out" to the concrete volume and finish area.

Shored Slabs should be listed on your takeoff by shoring height and floor number. This enables the person pricing the estimate to compare his costs on previous jobs for similarly shored slabs and to total the concrete material on each floor as mentioned under beams.

Hung Slabs are those hung from encased steel beams, including both composite and non-composite beams. When controlling costs during construction, it is difficult to separate the time and material used on the slab and the beams. For this reason, many estimators combine the cost of these, defining the beam size and slab thickness.

Domes and Pans can be placed on open plank centering on shoring or closed decking on shoring. Decide which method you're going to use and identify and estimate it in the same fashion as Shored Slabs. When computing the volume for slabs with domes or pans, figure the full depth of the slab (pan plus cover) and deduct the voids. The manufacturer provides a schedule of void volumes for the various sizes of domes or pans.

Stairs

When taking off cast-in-place stairs, a careful investigation should be made to ensure that all inserts such as railing anchors, nosings, and reinforcing steel have been included. Also, indicate on your Quantity Sheet any special treatment or finishes, which sometimes represent a considerable cost.

Curbs and Gutters

Concrete curbs and gutters should be identified and the quantities measured on the site work sheets of the plans listed on the Quantity Sheet of the estimate. Check off each location on the plans as you list the quantity on the Quantity Sheet.

Forms and Shoring

Slab on Grade

Slab on grade forming consists chiefly of edge forming. Pay particular attention to edge forms around "outs," changes in elevation, and changes in slab thickness. List edge forms by mm for each height.

Elevated Slab

In elevated slabs, unless your plyform is going to work out on exactly modular size, figure 15% overrun on the first use and 10% waste on each reuse. The support lumber will average 5.33×10^{-2} m^3 per m^2CA with shores placed 1200 mm O.C. under support beams, R031-050; Forms in Place, in *Means Building Construction Cost Data, Metric Version*. Hung slabs will require hangers in place of shores.

The following factors can affect the number of pans rented, the m^2 of forms you build, and the number of shores provided:

- Concrete pump or crane and bucket rental cost per pour.
- Placement rate of concrete for crew.
- Finishing rate for concrete finishers.
- Curing time before stripping and re-shore.
- Forming and stripping time.
- Number of form reuses.

Experience or consultation with an experienced superintendent will tell the estimator which combination of these six items will be the limiting factor on the m^2 of forms to be built and number of pans to be rented.

Stairs

Using the recommended method of estimating by slant length and width of stairs, the stringer, riser, and framing lumber figures to about 0.108 m³ per m²CA. The plyform can be figured with 10% waste. With careful stripping, it can be reused a maximum of four times.

Curbs and Gutters

Special metal curb and gutter forms can be purchased or rented. Except for special shapes or small quantities, these are more economical than building wooden forms and the concrete will require less finishing after stripping.

Placing

Some imagination and experience by the estimator can be used here. By leaving a spandrel beam out, can transit mix trucks enter a building and place the slab on grade by direct chute? By using particular sequence, can small wheeled pours be eliminated? As with placing of formed concrete, the site, location within the structure, sequence of pours, available equipment, and personal preference are all factors used in determining the placement method of concrete. See R033-090; Placing Ready Mixed Concrete, *Means Building Construction Cost Data, Metric Version*.

Write the method of placement on your Quantity Sheet for reference when you are pricing the estimate.

Slab on Grade

Placing slabs by direct chute because no equipment is involved can be done in large or small pours without appreciably affecting the unit placing cost. Usually the limiting factor is "How much can the finishing crew finish per day?"

Elevated Slab and Stairs

Most elevated slabs and stairs are placed by crane and bucket, concrete pump, or conveyor. Unless you are planning to price your estimate on historical unit costs as in *Means Building Construction Cost Data, Metric Version*, estimate the number of days this equipment will be needed and write this on your Quantity Sheet, i.e., 10 day, 13.6 metric ton crane and bucket.

Pricing Type II: Cast-in-Place Slabs and Other Finished Surfaces

Pricing of flatwork is comprised of combining the cost of forming reinforcing, placing, finishing, and concrete material. Using the Example of an Elevated Floor Slab turn to R031-060, Formwork Labor Hours in *Means Building Construction Cost Data, Metric Version*. Under Flat Slabs, the man-hours required for forming per m²CA are:

One Use		
Fabricate	0.39	MH
Erect and Strip	0.63	MH
Clean and Move	0.13	MH
Total	1.15	MH/m²CA

Now turn to 031-150-2000 and you will find that Crew C-2 is used to perform the task. Crew C-2 has 48 MH per day. At 1.15 MH per m²CA, the daily productivity is:

$(48/1.15) \times m^2CA = 41.74 \ m^2CA/day$

The labor cost for a single use (031-150-2000) is $26.50/m²CA. Multiple use labor hours are given in R031-060, *Means Building Construction Cost Data, Metric Version*.

Material

Concrete material pricing is as near as your telephone. The size of your project, proximity to the batch plant and competitive atmosphere at the time of the bid are all relevant factors. For an average price, use R033-060; Concrete Material Net Prices, R033-070; Ready Mix Material Prices or R033-080; Field Mix Concrete or turn to R033-122 and-126 in *Means Building Construction Cost Data, Metric Version*.

Type III: Precast Concrete and Cementitous Decks

Precast concrete and cementitous decks are the final concrete items to be discussed in our estimating methods. Other than site cast units such as tilt-up wall panels, most structural precast members are plant produced and purchased by the m or m^2 either delivered to the site or delivered and erected by the producer. For those items cast off the site, Division 034 has priced the delivery of material to the site and the erection cost separately.

Quantity Takeoff

Material

Define and quantify each type of member shown on the plans. The commonly used unit of measure for each member is given in Division 034 and 035 of *Means Building Construction Cost Data, Metric Version*.

Beams, Columns, Tees, and Wall Panels should be listed according to similar dimensions and load-bearing capacity. Beams and columns are generally listed by the m, whereas slabs and walls are listed by the m^2. A discussion of Tilt-Up Panels, Lift Slabs, Wall Panels, Prestressed Precast Concrete Structural Units, and Tensioned Members is covered in the back of *Means Building Construction Cost Data, Metric Version* in the Reference Numbers for Division 3.

Installation

Site cast units such as tilt-up walls are usually erected by the casting contractor. Write the quantity to be erected on your Quantity Sheet to be carried over to your pricing sheet.

Finishes

While studying the plans, indicate the finish, grouting, or topping shown or as defined in the specifications.

Pricing Type III: Precast Concrete and Cementitous Decks

Pricing of precast units is similar to structural steel or lumber or masonry—material delivered to the site plus installation.

- National average material prices are given in Division 034 and 035 of *Means Building Construction Cost Data, Metric Version*. Prices vary greatly in different regions of the country. Also, haul distance from the plant to a particular job site can affect the price. A quote from a local producer should be obtained if possible.
- Figure II.8 is the takeoff for the concrete construction of the sample project.

COST ANALYSIS

PROJECT: Repair Garage
LOCATION:
TAKE OFF BY: ABC
CLASSIFICATION: Division 3
ARCHITECT:
QUANTITIES BY: ABC
PRICES BY: As Shown
EXTENSIONS BY: DEF
CHECKED BY: GHI
SHEET NO. 4 of 15
ESTIMATE NO:
DATE: 1993

DESCRIPTION	SOURCE	QUANT	UNIT	MATERIAL COST	MATERIAL TOTAL	LABOR COST	LABOR TOTAL	EQUIPMENT COST	EQUIPMENT TOTAL	SUBCONTRACT COST	SUBCONTRACT TOTAL	TOTAL COST	TOTAL
Concrete													
Cont. Ftgs. Forms, 4 uses	031 158 0150	57	m2CA							29.00	1653		
Rebar, #10 to #15	032 107 0500	0.45	Met. Ton							1350	608		
Concrete, 21 MPa	033 126 0150	23	m3							69.00	1587		
Place & Vibrate	033 172 1900	23	m3							16.75	385		
Slab Reinf. - 6 x 6 W1.4 x W1.4	032 207 0010	409	m2							3.05	1247		
Slab Conc., 21 MPa	033 126 0150	40	m3							69.00	2760		
Place & Vibrate Slab	033 172 4300	40	m3							18.25	730		
Finish Slab	033 454 0250	372	m2							6.20	2306		
Precast Plank, 150 mm (R034-030) Delivered Cost	034 136 0050	46	m2	37.00	1702								
Erect w/ crew F-3	Crew F-3	1	Day			min.	947 min.		392				
Expansion Joints, 24 Ga. 114 mm High	031 132 0050	104	m							8.10	842		
Asphalt Seal	031 132 1100	104	m							3.00	312		
Division 3 Total					$1,702		$947		$392		$12,431		

Note: m2 = square meter m3 = cubic meter

Figure II.8

Division 4
Masonry

Masonry is generally priced by the unit or by the piece or per thousand units. The quantity of units is determined from square meter measurements of walls and partitions. The areas are converted into number of units by appropriate multipliers (see R042-500; Brick, Block & Mortar Quantities in *Means Building Construction Cost Data, Metric Edition*.)

The multipliers are a function of:

- Size of the masonry unit
- Thickness of mortar joints
- Pattern or coursing of the masonry unit
- Thickness of wall

The number of units per square meter is calculated, then multiplied by the wall area to determine the total quantity involved. For items installed in a course, the quantity per meter is calculated, then multiplied by the length involved.

Changes Required for Metric

For metrication of Division 4, size differences between Imperial size bricks and blocks and metric units are minor. Initially, productivity will suffer as masons and bricklayers make the transition from Imperial to metric units; however, this productivity loss should be minimal.

Brick

Conversion to metric will result in the following changes to brick dimensions. Standard brick will be changed to 90 x 57 x 190 mm. Mortar joints will change from 3/8" and 1/2" to 10 mm. Brick modules will change from 2' x 2' to 600 x 600 mm. Brick and mortar composition will remain the same. Five to ten of the approximately one hundred sizes of brick currently manufactured are within one millimeter of a metric brick.

BIA on Metric

The Brick Institute of America reports that brick in the metric system should conform to a 200 mm module for modular construction. Several existing sizes do conform to a 200 mm module when a 10 mm (roughly 3/8") mortar joint is used. See Figure II.9 for the actual dimensions in both inch and mm units of common brick sizes that fit the 200 mm module. These sizes fit modular metric dimensions with no physical changes.

Some brick that do not have modular metric dimensions in thickness or length do conform to the 200 mm coursing module. This vertical dimension is the most important for brick in metric construction, especially during the transition period to all-metric construction. Using a 10 mm mortar joint, the brick in Figure II.9 will course out as follows.

Modular and Standard H = 57 mm 3 courses: 200 mm (2-1/4″)

Engineer H = 70 mm 5 courses: 400 mm (2-3/4″)

Economy H = 90 mm 2 courses: 200 mm (3-1/2″)

Roman H = 40 mm 4 courses: 200 mm (1-5/8″)

One area that has not yet been addressed is the paving sector. Soft conversion of existing paver sizes is all that is currently required.

Concrete Block

To convert to metric, block sizes will change to 390 x 190 x 190 mm. The block module will change from 2′ x 2′ to 600 x 600 mm. Metric block sizes are 10 mm (3/8″) shorter in the long dimension from current block

UNIT NAME		ACTUAL DIMENSIONS inches	millimeters	METRIC MODULAR SIZE (mm)	EXAMPLE
Modular	T H L	3-1/2 2-1/4 7-1/2	89 57 190	90 57 190	
Standard	T H L	3-1/2 2-1/4 8	89 57 203	90 57 200	
Engineer Modular	T H L	3-1/2 2-3/4 7-1/2	89 70 190	90 70 190	
Economy Modular (Closure)	T H L	3-1/2 3-1/2 7-1/2	89 89 190	90 90 190	

Common Brick Sizes

(courtesy Brick Institute of America)

Figure II.9

sizes. The composition of both block and mortar will remain the same, although mortar joints will change from 1/2" to 10mm.

NCMA on Metric

The National Concrete Masonry Association (NCMA) appointed a Task Group on Metrication in 1990 in response to Executive Order 12770 requiring federal agencies to move forward to metric design of construction for federal projects. NCMA recognizes the significant cost of metrication for their industry, as well as the time required to carry out the necessary steps. The depressed construction market has caused some concern about the return on investment of metric changes.

International trade may be less of a factor for this product, since metric block units are produced only in the eastern provinces of Canada. Furthermore, concrete blocks are heavy and relatively fragile, thereby making them less attractive for shipping.

Block manufacturers and machinery suppliers anticipate a minimum three-year time frame to convert all production to metric. As an alternative to full compliance, consideration is being given to re-dimensioning to "soft" metric conversion, adjusting mortar thickness to meet location dimensions required by engineering or architectural plans.

Estimating Procedures

All masonry items that are to be priced should be represented by a notation on the quantity sheet. Before starting the quantity takeoff, the estimator should make notes based on the masonry specifications of all the items that will have to be estimated. These notes will include the kind of masonry units, the bonds, the mortar type, joint reinforcement, grout, miscellaneous installed items, scaffolding, cleaning, and so on.

The elevation shows the exterior wall, door and window openings. The plans show the number, area, and layout of interior walls. Sections, details, and the room finish schedule will serve to identify the different masonry units and show where they occur.

The estimator should make notes on the plans identifying each type of material, such as where face brick and concrete block backup are called out, as opposed to solid brick or exposed block.

Measurements

After the walls and partitions have been identified and analyzed, each kind or combination of masonry unit is measured and entered on the takeoff sheets. The measurements are total areas with all openings over one square meter deducted, and openings under one square meter ignored. The items are grouped in the order in which they would be constructed on the job.

- Walls below grade
- Partitions
- External walls
- Miscellaneous brickwork
- Chimneys

Walls below grade may be pointed on the exposed face and dampproofed on the earth side. The calculated area can also be used as a measurement of the dampproofing. Scaffolding is required for walls over 1200 mm high. Partitions below grade are no different from other partitions, so they do not have to be kept separate.

External wall areas are the product of the wall length multiplied by the actual wall height taken from the section detail. A small scale sketch of the building posted on a nearby wall is an important visual aid for future reference.

It should be remembered that exterior measurements of perimeter give an overlap of four times the thickness of the walls. This is accepted accuracy because corners tend to require more cutting and waste.

Mortar

The estimator will often have no choice in determining the mortar proportion since the specifications indicate the type to be used. The mortar is very important to the structural integrity of the wall—it is false economy to alter a specification to save money on a mortar mix. The following chart describes the uses and strengths of various types of mortar mixes.

Brick Mortar Mixes*

Type	Portland Cement	Hydrated Lime	Sand (max.)	**Strength	Use
M	1	1/4	3-3/4	High	General use where high strength is required, especially good compressive strength; work that is below grade and in contact with earth.
S	1	1/2	4-1/2	High	Okay for general use, especially good where high lateral strength is desired.
N	1	1	6	Medium	General use when masonry is exposed above grade; best to use when high compressive and lateral strengths are not required.
O	1	2	9	Low	Do not use when masonry is exposed to severe weathering; acceptable for non-load-bearing walls of solid units and interior non-load-bearing partitions of hollow units.

The water used should be of the quality of drinking water. Use as much water as is needed to bring the mix to a suitably plastic and workable state.

**The sand should be damp and loose. A general rule for sand content is that it should not be less than 2.25 or more than 3 times the sum of the cement and lime volumes.*

Bonding

Structural bonds may be accomplished in three ways:

- By the overlapping (interlocking) of masonry units.
- By the use of metal ties embedded in connecting joints.
- By the addition of grout between adjacent wythes of masonry.

Frequently, structural bonds, such as English or Flemish or variations of these, may be used to create patterns in the face of the wall. However, in the strict sense of the term, *pattern* refers to the change or varied arrangement of the brick. When using any of these coursing patterns, an adjustment to the total number of face bricks must be taken into account if the bricks' lengths are twice their bed depths. The sketches in Figure II.10 show some of these traditional bond patterns.

TRADITIONAL BONDING PATTERNS

Running or Stretcher Bond — The face bricks are all stretchers and are tied to the backing by metal or reinforcing.
Waste: 5%

Common or American Bond — Every sixth course of stretcher bond is usually made a header course.
Waste: 4%

Flemish Bond — Each course has alternate headers and stretchers with the alternate headers centered over the stretcher.
Waste: 3 to 5%

English Bond — Consists of alternated headers and stretchers with the vertical joints in the header or stretcher aligning or breaking over each other.
Waste: 8 to 15%

Stack Bond — Has no overlapping of units since all vertical joints are aligned. Usually this pattern is bonded to the backing with rigid steel ties.
Waste: 3%

English Cross or Dutch Bond — Built up of interlocking crosses. This wall consists of two headers and a stretcher forming a cross.
Waste: 8%

Figure II.11 illustrates brick dimensions and shapes, positions, and surfaces, as well as common mortar and masonry joints.

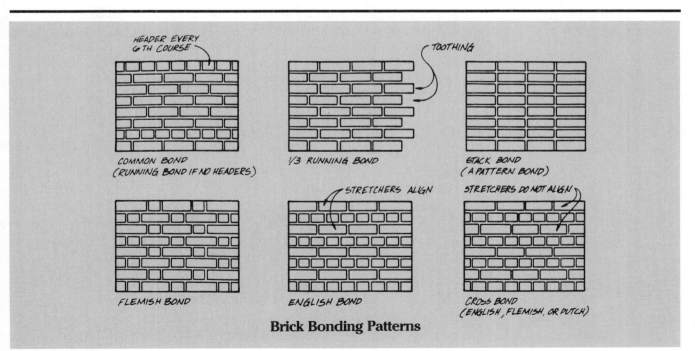

Brick Bonding Patterns

Figure II.10

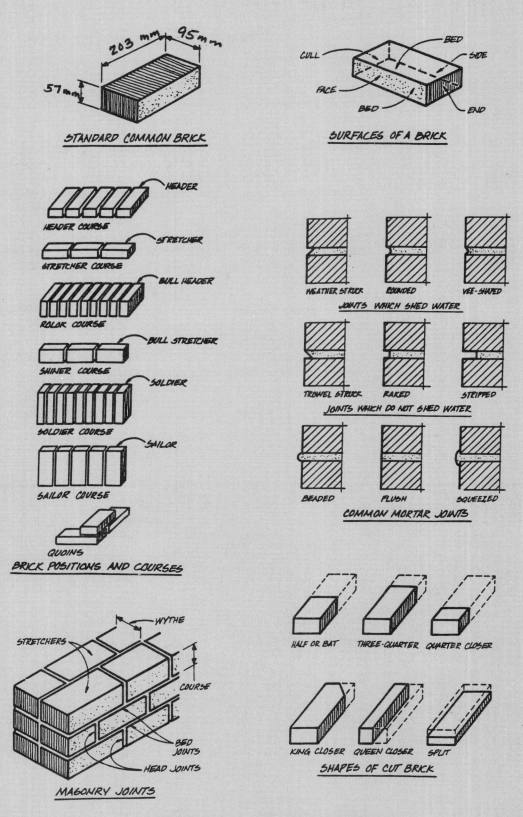

Brick Dimensions, Shapes, and Joints

Figure II.11

Loose Lintels

The following is a typical lintel specification:

All loose lintels not shown on the structural drawings shall be supplied under this section in accordance with the following schedule: Supply one angle for each 102 mm thickness of masonry. Longer legs shall be set vertically. Install lintel over each masonry opening.

Concrete Block Masonry

Interior partition heights are dependent upon whether they stop at a course above the suspended ceiling line or extend to the bottom of the floor slab above. When partitions are the same height but of different thicknesses, the most efficient takeoff method is to measure all the partitions with the same thicknesses before multiplying by height. Some estimators take off the partitions by measuring all horizontal distances, working from the top of the plan to the bottom and then adding the vertical distance, working from the left-hand side to the right. Figure II.12 shows a standard concrete block with metric dimensions, and various concrete shapes.

Often there are many combinations of masonry materials on a job. The simple example of a concrete block wall shows some of the various factors an estimator must consider.

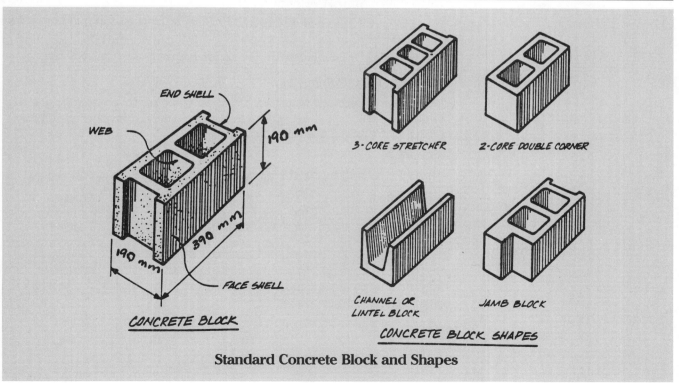

Standard Concrete Block and Shapes

Figure II.12

Calculating Wall Area and Number of Block in Running Bond

Multiply the length of the wall by the height; from this total area deduct all openings larger than 0.2 square meters (smaller openings should be ignored) and count the areas of the openings.

Multiply the adjusted m² area by 12.1* to get the blocks per square meter. This will give the number of 203 mm x 406 mm C.M.U.

$$*\text{Note:} \frac{1000k \text{ mm}^2/\text{m}^2}{203 \text{ mm} \times 406 \text{ mm}} = 12.1 \frac{\text{blocks}}{\text{m}^2}$$

- At all abrupt changes in wall height;
- At all changes in wall thickness, such as pipe chases, and adjacent to columns and pilasters;
- Above joints in foundations and floors;
- Below joints in roofs and floors that bear on walls;
- At a distance of not over 1/2 the allowable joint spacing from bonded intersections of corners; and
- At one or both sides of all door and window openings unless other control measures, such as joint reinforcement or bond beams, are used.

Lintels over openings may be steel angles, precast units, or bond beams of 203 mm x 406 mm blocks. If bond beam blocks are to be replaced by lintel blocks that are 203 mm x 203 mm, add two lintel blocks for each bond beam block.

Concrete Block Wall Ties

Metal wall ties that are specified at a certain horizontal and vertical spacing are figured by multiplying the vertical spacing in mm and horizontal spacing in mm, then dividing by square meters.

Example: 406 mm O.C. Vertical $\frac{406 \text{ mm} \times 914 \text{ mm}}{1000 \times 1000 \text{ mm}^2/\text{m}^2} = 0.37 \text{ m}^2/\text{tie}$

914 mm O.C. Horizontal

By taking the reciprocal of the area of the wall/tie, you get the number of ties per square meter of wall.

Example: 1/0.37 = 2.7 ties/m²   2.7 x 100 m² = 270 ties/100 m²

Concrete Block Reinforcing

Horizontal reinforcing is usually specified on some multiple vertical spacing of 203 mm. To calculate the length of reinforcing, take the number of 203 mm courses and subtract one from the number.

Example:

3 m high wall = 3 m / 0.203 m/course = 15 courses − 1 = 14 courses

	then divide by					
		1	for	203 mm On Center	=	14 courses
		2	for	406 mm On Center	=	7 courses
		3	for	610 mm On Center	=	4 courses
		4	for	812 mm On Center	=	3 courses
		5	for	1015 mm On Center	=	2 courses
		6	for	1218 mm On Center	=	2 courses

Then multiply by the length of the wall. 6 m wall = 6 m x 14 = 84 m for reinforcing 203 mm on center.

Vertical reinforcing is usually specified on some multiple horizontal spacing of 203 mm with minimum reinforcement @ 1220 mm O.C. and maximum reinforcement @ 410 mm O.C. The following table can be used to calculate the number of vertical pieces of reinforcement required. When the number of pieces has been determined, simply multiply the quantity by the wall height to determine the total vertical steel required.

Specified Spacing	Horizontal Length Multipliers
410 mm O.C.	0.75
610 mm O.C.	0.50
810 mm O.C.	0.375
1220 mm O.C.	0.25

The estimator must be alert to the added steel that may be required around openings, at corners, wall to concrete slab connections, laps, and dowels.

Miscellaneous Items of Work Affecting Brick and Block Masonry

When estimating solid or hollow unit masonry, the estimator must give consideration to many items: anchor bolts, ledgers, fasteners, waterproofing, solid grouting, continuous inspection, metal frames, sealants, reglets and flashings, insulation, and slurry coating. The miscellaneous items associated with masonry are often F.O.B. items by a fabricator and, with installation, the responsibility of the Bricklayer or the General Contractor.

Figure II.13 is the takeoff form for the masonry construction on the sample project.

COST ANALYSIS

PROJECT: Repair Garage
LOCATION:
TAKE OFF BY: ABC
CLASSIFICATION: Division 4
ARCHITECT:
QUANTITIES BY: ABC
PRICES BY: As Shown
EXTENSIONS BY: DEF
CHECKED BY: GHI
SHEET NO. 5 of 15
ESTIMATE NO:
DATE: 1993

DESCRIPTION	SOURCE			QUANT	UNIT	MATERIAL COST	MATERIAL TOTAL	LABOR COST	LABOR TOTAL	EQUIPMENT COST	EQUIPMENT TOTAL	SUBCONTRACT COST	SUBCONTRACT TOTAL	TOTAL COST	TOTAL TOTAL
Masonry															
C.M.U. Foundation, 305 mm	042	320	0350	74	m2	24.50	1813	28.00	2072						
Exterior Block w/ inserts, 305 mm	042	310	0300	391	m2	32.00	12512	33.50	13099						
Interior Partition, 203 mm, L.W.	042	232	1200	37	m2	17.85	660	24.50	907						
Interior Partition, 102 mm, L.W.	042	232	1100	37	m2	10.25	379	21.50	796						
Division 4 Total							$15,365		$16,873						

Note: m2 = square meter

Figure II.13

Division 5
Metals

The metals portion of a building project, and the corresponding estimate should be broken down into basic components: structural metals, metal joists, metal decks, miscellaneous metals, ornamental metals, expansion control and fasteners. The items in a building project are shown on many different sheets of the drawings and may or may not be thoroughly listed in the specifications. This is especially true of the miscellaneous metals that are listed under Division 5. A complete and detailed review of the construction documents is therefore necessary, noting all items and requirements. Most structural steel work is subcontracted to specialty fabricators and erectors. However, the estimator for the general contractor may perform a takeoff to assure that all specific work is included. Pricing based on total weight (tonnage) can be used to verify subcontractor prices.

Changes Required for Metric

Structural Steel

Metrication of construction will require a number of changes for structural steel. Section designations will be in millimeters rather than inches. Pounds per foot will be translated to kilograms per meter. Bolt diameters will be in metric, and threads will be based on ASTM A325M and A490M. Nominal cross section designations may change to metric, but their actual dimensions will remain the same.

On repair and remodeling construction projects, under Division 5, there will be a need for transition pieces or special connection details to ensure that the various existing Imperial and new metric-sized beams, columns, and bracings are adequately connected to each other.

AISC on Metric

The American Institute of Steel Construction (AISC) has performed a soft conversion of all structural steel shapes. A detailed report is available from that organization. Their expectation is that within five years, selected shapes will be manufactured to hard metric dimensions.

Estimating Procedures

Structural Steel

- Specification information includes type of steel, connections, bolts, cleaning methods, paint required and number of coats, type and number of detail drawings to be submitted, type of testing in field and shop, and whether the steel will be architecturally exposed.
- Drawing information includes special notes; lintel schedule; size, weight, and location of each piece of steel; anchor bolt locations; special fabrication details; special weldment; and special fabrication.
- The estimator must give consideration to mill deliveries, warehouse availability, type of fabrication required, fabricator capability, and availability. Delivery dates, sequence of deliveries, size of pieces to be delivered, and site storage area are of equal importance. Engineering and detailing, underwriter labels, and field dimensions must all be included in the estimated cost, but usually are not defined in the specifications or detailed on the drawings. These are often provided by the fabricator ... make certain you verify.

Steel Joists

- Specification information includes types of steel, design criteria, and type of paint.
- Drawing information includes type of joists, size and length, end bearing conditions, tie joist locations, bridging type and size, header types and sizes, top and bottom chord extensions, and ceiling extensions.
- The estimator must consider the various types and sizes, special attachments, special paint and preparation, deliveries, and other variables that must be included in the estimate but are not called out.

Steel Deck

- Specification information includes type of steel, design criteria, and finishes.
- Drawing information includes gauge, depth, rib spacing, finishes, sumps, and closures required.
- Again, it is important to consider the quantity, availability, and delivery schedule of the types of deck specified, as well as site storage, special requirements, and miscellaneous items necessary to complete the work.

Miscellaneous Metals

- Specification information includes the description of many required items and various materials and finishes for the items. The miscellaneous section is responsible for all miscellaneous items required by another division (usually 3, 4, 6, 15, and 16) but not specified in that division.
- Drawing information may include details of the miscellaneous metal required, but often notes on the drawings are the only source of information available to the estimator.
- Because the miscellaneous metals are difficult to identify, it is important to carefully inspect all drawings and add all necessary items not noted or specified. They are usually well hidden in the construction. Determination and experience are sometimes the only keys to finding these items.

Ornamental Iron

- Specification information includes a description of most of the ornamental items required, materials and finishes, and type of details required of the fabricator for approval.

- Drawing information usually includes notes and details of items required, and shows the relationship of ornamental metal to other work.
- Consider the material availability, finishes required, protection of finished products, installation procedures, coordination with other trades and, finally, items not shown but needed to complete the installation.

Erection

- Specification information includes items to be erected, criteria required to erect various items, special connections, type of welds or bolts, tolerances required, and special erection procedures.
- Drawing information includes items to be erected, size, quantity, location, and supporting structures.
- Erection costs cannot be determined without giving consideration to site storage, site access, type and size of equipment required, delivery schedules, and erection sequence.

If the prime contractor applies the information discussed in this chapter, the sub-bids received for Division 5 can be evaluated more closely. From the evaluation and comparison with subcontractor proposals, items that were omitted can be identified and quickly priced using *Means Building Construction Cost Data, Metric Version*.

Figure II.14 is the quantity takeoff form for the structural steel part of the sample project.

COST ANALYSIS

PROJECT: Repair Garage
LOCATION:
TAKE OFF BY: ABC
CLASSIFICATION: Division 5
ARCHITECT:
QUANTITIES BY: ABC
PRICES BY: DEF
EXTENSIONS BY: DEF
CHECKED BY: GHI
SHEET NO. 6 of 15
ESTIMATE NO:
DATE: 1993

DESCRIPTION	SOURCE			QUANT	UNIT	MATERIAL COST	MATERIAL TOTAL	LABOR COST	LABOR TOTAL	EQUIPMENT COST	EQUIPMENT TOTAL	SUBCONTRACT COST	SUBCONTRACT TOTAL	TOTAL COST	TOTAL TOTAL
Metals															
Frames, Lintel, & Channels	051	230	0600	3851	kg	1.65	6354	0.53	2041	0.05	193				
Open Web Joists, 12 m Span	052	110	0300	5.9	Met. Ton	660.00	3894	133.00	785	74.00	437				
Metal Deck, 38 mm 22 Ga. Open	053	104	2100	372	m2	7.85	2920	2.03	755	0.18	67				
Steel Stair, 1.1 m W, Grating Tread & Pipe Railing	055	104	0020	14	Riser	70.00	980	24.00	336	2.09	29				
Division 5 Total							$14,148		$3,917		$725				

Note: m2 = square meter

Figure II.14

Division 6
Wood and Plastics

While the use of wood as a building product in commercial buildings was greatly reduced because of stringent fire codes, this trend is reversing itself in the small condominium office buildings being built in various areas of the country. New technology in the assembly of wood products has decreased the cost of the finished product. Wood is an economical, reproducible product that should be recognized for its strength, durability, workability, and finished appearance. Material prices for lumber fluctuate more and with greater frequency than any other building material. For this reason, when your material list is complete for a project, obtain current, local prices on the lumber.

Changes Required for Metric

Division 6 will be affected by metrication in the following ways. Plywood sheet sizes will change from the current 4' x 8' size to a new hard metric size of 1200 mm x 2400 mm. Spacing will change from 16" and 24" O.C. to 400 mm and 600 mm respectively. The 2" x 4" may keep its nominal size; however, it will be specified as 50 mm x 100 mm.

Wood building products are being soft-converted from their nominal or exact dimensions to metric equivalents. While the actual dimensions of framing lumber are not expected to change, studs can be spaced at 400 mm (15-3/4") instead of 16". The width of batt insulation installed between studs need not be changed; it will simply be more of a "friction fit." Plywood may be converted from 8'-0" to 2400 mm, 10'-0" to 3000 mm.

Wood Product Associations on Metric

The American Forest and Paper Association (AFPA) reports that, in general, metric conversion of lumber and wood building products will be soft for the short-term. Particleboard and nonstructural panels may lead the way in conversion to hard metric dimensions, but even these will be soft converted until industry practitioners commit to converting to new modules for building, such as 400 millimeters on center instead of 16" O.C. AFPA indicates that manufacturers of lumber and wood building products are ready to respond with products in metric sizes dictated by the market, as trade groups commit to the change.

The Western Wood Products Association (WWPA) has provided its members with soft conversions of standard lumber dimensions on basic wood building products and on a wide range of incremental sizes (see Figure II.15). WWPA, like AFPA, anticipates the use of predominantly soft metric conversions in the next few years. WWPA notes that while mills are capable of producing, and do produce, materials in metric dimensions with no requirement or delays for re-tooling, economic feasibility can be a major issue because of the relatively small orders to date in metric.

The American Plywood Association (APA) currently prints all of its literature in metric for use overseas. APA indicates a soft conversion of 4' x 8' plywood sheets to 1200 x 2400 mm as a rough (nominal) approximation. Actual (soft-converted) metric dimensions are 1220 mm x 2440 mm. APA notes that many Canadian mills produce plywood sheets in hard metric dimensions.

Estimating Procedures

Rough Carpentry

In metric dimensions, lumber is usually estimated in cubic meters and purchased in cubic meter quantities.

The Quantity Survey should indicate species, grade, and any type of wood preservative or fire retardant specified or required by code.

Sills, posts, and girders used in subfloor framing should be taken off by length and quantity. The standard lengths available vary by species and dimensions, and cut-offs often are used for blocking. Careful selection of lengths will affect the waste factor required.

Floor joists, shown or specified by size and spacing, should be taken off by length and quantity required. Do not forget double joists under partitions, headers at openings, overhangs, laps at bearings, and blocking or bridging.

The studs required are noted on the drawings by certain spacings, usually 400 or 600 mm O.C., with the size of stud given. Taking the linear meters of partitions that are similar in stud size, height, and spacing, and dividing the linear meters by the spacing will give the number of studs required. Added studs required for corners and intersecting partitions must be taken off separately.

The number and size of openings are important. Even though there are no studs in these areas, headers, subsills, king studs, trimmers, cripples, and knee studs must be taken off. Remember that a sole plate (sometimes double) and double top plate are necessary. Where bracing and fireblocking are noted, you should indicate the type and quantity.

Ceiling joists are similar to floor joists but will usually vary in direction. One must be alert for soffits and suspended ceilings. Ledgers may be a part of the ceiling joist system. In a flat roof system the rafters may be called rafter joists and shown as a ceiling system.

Roof rafters vary with the different types of roofs. A hip and valley, because of its complexity, has a greater material waste factor than most other roof types. Although the straight gable, gambrel and mansard type roofs are not as complicated, care should be taken to ensure a good material takeoff. Roof pitches, overhangs, and soffit framing all affect the quantity of material and costs. Do not scale rafters from the plan; always use the elevations.

Roof trusses are usually furnished and delivered to the job site by the truss fabricator. Make certain that the fabricator furnishes shop drawings and that

In January, 1989, regulations went into effect requiring that all U.S. lumber export volumes be reported in metric terms. The new regulations are designed to bring the U.S. in line with the statistical reporting methods used by other exporting and importing nations. The following formulas and tables set forth methods to convert commonly exported sizes to metric volumes.

Quick Formulas:

1. If lumber is full sawn, or footage is computed on actual sizes, multiply the board footage expressed in thousands of board feet (MBF) by 2.358 to find cubic meters:

   $$\text{MBF} \times 2.358 = m^3$$

2. If lumber footage is based on nominal sizes, divide the actual cross section (thickness x width) by the nominal cross section (thickness x width), then multiply by 2.358. Multiply the total board footage (MBF) by this figure to find the total cubic meters:

   $$\frac{\text{Actual cross section}}{\text{Nominal cross section}} \times 2.358 \times \text{MBF} = m^3$$

Example: 2x4 dry S4S

$$\frac{1\frac{1}{2}'' \times 3\frac{1}{2}''}{2'' \times 4''} \times 2.358 \times \text{MBF} = m^3$$

3. If the lumber is trimmed to a specific length, but billed on an even foot basis, as in precision end-trimmed studs, then account for this difference by multiplying the total cubic meters (found by either of the above formulas or the chart) by an additional factor. To find this factor, divide the actual trim length by the nominal length, then multiply the total cubic meters by this number:

   $$\frac{\text{Actual trim}}{\text{Nominal length}} \times m^3 = \text{total } m^3$$

Quick Lumber Conversions:

Inches x 25.4 = Millimeters

Millimeters x .03937 = Inches

Feet x .3048 = Meters

Meters x 3.28 = Feet

Cubic Feet x 0.02832 = Cubic Meters

Cubic Meters x 35.314 = Cubic Feet

Quick Fraction Conversions:

Fraction	Decimal	Millimeters
$\frac{1}{16}$	.0625	1.59
$\frac{1}{8}$	.1250	3.18
$\frac{3}{16}$	.1875	4.76
$\frac{1}{4}$	.2500	6.35
$\frac{5}{16}$	.3125	7.94
$\frac{3}{8}$	.3750	9.53
$\frac{7}{16}$	.4375	11.11
$\frac{1}{2}$	.5000	12.70
$\frac{9}{16}$	.5625	14.29
$\frac{5}{8}$	.6250	15.88
$\frac{11}{16}$	.6875	17.46
$\frac{3}{4}$	.7500	19.05
$\frac{25}{32}$	.7813	19.84
$\frac{13}{16}$	.8125	20.64
$\frac{7}{8}$	.8750	22.23
$\frac{15}{16}$	.9375	23.81
1.0	1.0000	25.40

Using the tables below:

If lumber footage is figured on nominal size, find the nominal measure in the chart below, under the corresponding heading (i.e., Dry, Green, etc.), then multiply the board footage expressed in thousands of board feet (MBF) by the corresponding factor to find cubic meters.

Note: This m^3/MBF factor is derived by $\frac{\text{Actual cross section}}{\text{Nominal cross section}} \times 2.358$

Dry S4S

Nominal Size	Actual Size	m^3/MBF
1x2	$\frac{3}{4} \times 1\frac{1}{2}$	1.326
1x3	$\frac{3}{4} \times 2\frac{1}{2}$	1.474
1x4	$\frac{3}{4} \times 3\frac{1}{2}$	1.547
1x6	$\frac{3}{4} \times 5\frac{1}{2}$	1.621
1x8	$\frac{3}{4} \times 7\frac{1}{4}$	1.603
1x10	$\frac{3}{4} \times 9\frac{1}{4}$	1.636
1x12	$\frac{3}{4} \times 11\frac{1}{4}$	1.658
2x3	$1\frac{1}{2} \times 2\frac{1}{2}$	1.474
2x4	$1\frac{1}{2} \times 3\frac{1}{2}$	1.547
2x6	$1\frac{1}{2} \times 5\frac{1}{2}$	1.621
2x8	$1\frac{1}{2} \times 7\frac{1}{4}$	1.603
2x10	$1\frac{1}{2} \times 9\frac{1}{4}$	1.636
2x12	$1\frac{1}{2} \times 11\frac{1}{4}$	1.658

Baby Squares

Nominal Size	Actual Size	m^3/MBF
$3\frac{9}{16} \times 3\frac{9}{16}$	90mm x 90mm 3.543 x 3.543	2.358
$4\frac{1}{8} \times 4\frac{1}{8}$	105mm x 105mm 4.134 x 4.134	2.358

Scantlings Dry

Nominal Size	Actual Size	m^3/MBF
$\frac{6}{4} \times 3$	35mm x 70mm 1.378 x 2.756	1.990
$\frac{6}{4} \times 4$	35mm x 90mm 1.378 x 3.543	1.919
2x3	45mm x 70mm 1.772 x 2.756	1.919
2x4	45mm x 90mm 1.772 x 3.543	1.850

Green S4S

Nominal Size	Actual Size	m^3/MBF
1x4	$\frac{25}{32} \times 3\frac{9}{16}$	1.641
1x6	$\frac{25}{32} \times 5\frac{5}{8}$	1.727
2x3	$1\frac{9}{16} \times 2\frac{9}{16}$	1.574
2x4	$1\frac{9}{16} \times 3\frac{9}{16}$	1.641
2x6	$1\frac{9}{16} \times 5\frac{5}{8}$	1.727
2x8	$1\frac{9}{16} \times 7\frac{1}{2}$	1.727
2x10	$1\frac{9}{16} \times 9\frac{1}{2}$	1.750
2x12	$1\frac{9}{16} \times 11\frac{1}{2}$	1.765
3x4	$2\frac{9}{16} \times 3\frac{9}{16}$	1.794
3x6	$2\frac{9}{16} \times 5\frac{5}{8}$	1.888
3x8	$2\frac{9}{16} \times 7\frac{1}{2}$	1.888
3x10	$2\frac{9}{16} \times 9\frac{1}{2}$	1.913
3x12	$2\frac{9}{16} \times 11\frac{1}{2}$	1.930
4x4	$3\frac{9}{16} \times 3\frac{9}{16}$	1.870
4x6	$3\frac{9}{16} \times 5\frac{5}{8}$	1.969
4x8	$3\frac{9}{16} \times 7\frac{1}{2}$	1.969
4x10	$3\frac{9}{16} \times 9\frac{1}{2}$	1.995
4x12	$3\frac{9}{16} \times 11\frac{1}{2}$	2.013

Scantlings Green

Nominal Size	Actual Size	m^3/MBF
$1\frac{1}{2} \times 3$	$1\frac{3}{8} \times 2\frac{7}{8}$	2.071
$1\frac{1}{2} \times 4$	$1\frac{3}{8} \times 3\frac{7}{8}$	2.094
2x3	$1\frac{7}{8} \times 2\frac{7}{8}$	2.119
2x4	$1\frac{7}{8} \times 3\frac{7}{8}$	2.142
2x6	$1\frac{7}{8} \times 5\frac{7}{8}$	2.165
2x8	$1\frac{7}{8} \times 7\frac{7}{8}$	2.176
2x10	$1\frac{7}{8} \times 9\frac{7}{8}$	2.183
2x12	$1\frac{7}{8} \times 11\frac{7}{8}$	2.188

(courtesy Western Wood Products Association)

Figure II.15

erection costs are included in the costs. Trusses that are architecturally exposed usually are more expensive to fabricate and erect.

Roof decks of various woods, tongue and groove planks, or laminated are usually 50 mm to 100 mm thick and used mostly with glued laminated beams or heavy timber.

The square meter method is used to determine quantities with consideration given to roof pitches and nonmodular areas for waste. The materials are purchased by cubic meter measurement, and the conversion from square meter to cubic meter must allow for net sizes.

Sheathing on walls can be plywood of different grades and thicknesses, wallboard, or solid boards nailed directly to the studs. Plywood can be applied with the grain vertical, horizontal, or diagonal to the studding. Solid boards are usually nailed diagonally, but when lateral forces are not considered, they can be applied horizontally. Wallboards can be installed either horizontally or vertically, depending upon wall height and board dimensions. In estimating quantities of plywood or wallboard sheathing, the area to be covered in square meter divided by sheet size will indicate the number of sheets required. Diagonal or non-modular areas will create waste and must be considered in the estimate. For solid board sheathing, add 15% to 20% more material to the takeoff when using tongue and groove sheathing as compared with square edge. For diagonal application of boards, plywood, or wallboard, include an additional 10% to 15% material waste factor.

Subfloors can be CDX plywood; the thickness depends on the load and span, whether solid boards are laid diagonally or perpendicular to the joists, or tongue and groove planks are used. The quantity takeoff is similar to that used for sheathing noted above.

Stressed skin plywood includes prefabricated roof panels with or without bottom skin, floor panels, curved roof panels with or without bottom skin or tie rods, folded plate roof with intermediate rafters, all requiring the square meter quantity method.

Structural joists are prefabricated wood flanges with plywood webs or tubular steel open webs. This type of joist is spaced in accordance with the load and requires bridging and blocking supplied by the fabricator. The quantity takeoff should include the following: type, number required, length, spacing, end bearing conditions, number of rows and length of bridging, and blocking.

Grounds are normally 25 x 50 mm wood strips used for case work or plaster and the quantities are estimated in meters.

Furring is 25 x 50 mm or 25 x 75 mm wood strips generally fastened to masonry or concrete walls so that wall coverings may be attached thereto. Furring is also used on the underside of the ceiling joist to fasten the ceiling to. Quantities are estimated in meters.

Stair Stringers are usually of select stock.

Plans and specifications should be carefully checked for lumber and plywood treatments against decay, warping, fire, and insects. Depending upon the treatment, this can sometimes double the cost of the material. Some of the most common wood treatments and current prices are listed in Division 063 of *Building Construction Cost Data, Metric Version*.

Finish Carpentry and Laminated Construction

With the ever-increasing cost of job site labor and the improved technology of mass production, an entire room or half a house is sometimes hoisted into place with no finish carpentry required. One of the results of this trend is the almost universal use of prefabricated cabinets, counter tops, and shelves. Most cabinets, counters, and shelves can be taken off in items of stated dimension. Those that obviously cannot be preassembled should be taken off separately. Counter cutouts and radiator or air conditioner cutouts should be shown as labor items.

Carefully check the plans for nonstructural decorative beams and columns that were not included in the rough carpentry.

Millwork frequently has similar dimensions for at least two to four items. It's safer to list each item separately on your Quantity Sheet. All of these except door and window trim are listed by the linear meter. Standard door and window trim are listed per set. Exterior trim other than door and window trim should be taken off with the siding as the details and dimensions are interrelated. The common use of prehung doors and windows makes it convenient to take off this trim with the doors and windows.

Paneling is taken off by the square meter by type. Be sure to list any millwork that would show up on the details. Also include a percentage for waste. Ply siding and exterior trim are taken off by the square meter and linear meter, respectively. Be sure to allow an amount for waste.

There are as many ways of taking off stairs as there are different types of stairs. The important thing is, **don't leave them out**. Division 6 and *Wood Stair, Residential* of *Means Building Construction Cost Data, Metric Version* list many combinations of stairs and stair parts which should fit most any detail and specification.

The takeoff units must be adapted to the system: Square meter — Floor, Linear meter — Members, or Cubic meter — Lumber. As the members are factory fabricated, the plans and specifications must be submitted to a fabricator for takeoff and pricing.

Figure II.16 is the quantity takeoff for the wood and plastics portion of the sample project.

COST ANALYSIS

PROJECT: Repair Garage	CLASSIFICATION: Division 6	SHEET NO. 7 of 15	
LOCATION:	ARCHITECT:	ESTIMATE NO:	
TAKE OFF BY: ABC	QUANTITIES BY: ABC	PRICES BY: As Shown	DATE: 1993
	EXTENSIONS BY: DEF	CHECKED BY: GHI	

DESCRIPTION	SOURCE			QUANT	UNIT	MATERIAL		LABOR		EQUIPMENT		SUBCONTRACT		TOTAL
						COST	TOTAL	COST	TOTAL	COST	TOTAL	COST	TOTAL	
Woods & Plastics														
Misc. Blocking, 50 mm x 200 mm	061	102	2780	1.2	m3	180.00	216	375.00	450	16.65	20			
Roof Cants, Split 100 mm x 100 mm	061	120	7940	0.7	m3	229.00	160	184.00	129	8.10	6			
Facia, Precast & Slab, 25 mm x 200 mm	062	220	3370	12	m	1.15	14	2.72	33					
Counter Top, 22 mm Thick Square Edge, Plastic Face w/ Splash	062	408	1000	2.4	m	82.00	197	20.50	49					
Division 6 Total							$587		$661		$26			

Note: m3 = cubic meter

Figure II.16

Division 7

Thermal and Moisture Protection

This division includes all types of materials for sealing the outside of a building against moisture and air infiltration, plus the insulation and accessories used in connection with them. The estimator's principal concern is to determine the most probable places where these materials will be found in or on a building, the best methods of taking off quantities, and the easiest, most accurate way of estimating the installed cost of each material.

Changes Required for Metric

When Division 7 is fully metricated, rigid insulation board will be changed from 8' x 4' to 2400 mm x 1200 mm. Most material thicknesses within this division will be soft converted.

Insulation

Initially, the actual dimensions of insulation are not expected to change. Conversion to metric will be soft. A potential change in stud spacing from 16" to 400 mm (15-3/4") would simply result in more of a "friction fit." Thickness will remain the same, as will thermal ratings.

NICA on Metric

The National Insulation Contractors Association (NICA) reports no hard conversion for the short term, noting the significance of remodeling (non-metric) construction for their products.

Roofing

Metric dimensions for roofing will be: meter or millimeter for length, square meter for area, and millimeter/meter for slope.

Some roofing products, such as asphalt shingles, are readily available in hard metric sizes. Higher-end products, such as some asphalt shingles, are being offered in metric sizes earlier than lower-end materials, such as roll roofing. The availability of metric shingles also seems to vary regionally within the U.S. Canadian-made shingles have been sold and used in the U.S. for many years. They have been, and will continue to be sold in bundles representing a square (100 s.f.) of roof coverage. The standard metric shingle is 13-1/4" x 39-3/8", or 337 mm x 1000 mm. There is a variety of sizes (usually larger than the standard U.S. shingles) for "architectural" specifications. Regardless of unit size, these too will be specified in squares. Metal roofing will be soft-converted for the next several years.

Flashing
Metrication will require that the sheet metal designation change from "gauge" to millimeters. Actual thickness will remain the same.

Siding
According to the GSA Design and Construction Division, curtain wall systems are easily produced in and are, in fact, already available in hard metric sizes.

Wood siding products are undergoing a soft metric conversion for the short term. More information on wood building products can be found in Division 6, "Changes Required for Metric".

Estimating Procedures

Dampproofing usually consists of one or two bituminous coatings applied to foundation walls from about the finished grade line to the bottom of the footings. The areas involved are calculated from the total height of the dampproofing and the length of the wall. After the separate areas are figured and added together to provide a total square meter area, a unit cost per square meter can be selected for the type of material, the number of coats, and the method of application specified for the building.

Waterproofing of elastomeric sheets or membranes is estimated on the same basis as dampproofing, with two basic exceptions—the installed unit costs for the elastomeric sheets do not include bonding adhesive or splicing tape, which must be figured as an additional cost, and the membrane waterproofing on slabs must be estimated separately from the higher cost installation on walls. In all cases, unit costs are per square meter of covered surface and should be estimated on this basis. Waterproofing also includes clay or cement coatings, which expand on contact with water and prevent its further passage through the masonry wall. In the case of the clay-based Bentonite, the material is supplied in panels for direct application or in bulk for troweling on the wet side of the wall or slab.

Metallic coating material is applied to floors or walls, usually on the interior or dry side and after the masonry surface has been prepared by chipping or hacking for bonding to the new material.

The unit cost per square meter for these materials depends on the thickness of the material, the position of the area to be covered, and the preparation required. In many places where these materials are applied, access may be difficult and under the control of others. The estimator should make an allowance for delays caused by this type of problem.

Caulking and sealants are usually applied on the exterior of the building except for special conditions on the interior. In most cases, caulking and sealing is done to prevent water and/or air from entering a building. Therefore, at door and window frames and in other places where different materials meet, caulking and sealing is usually required. To estimate the installed cost of this type of material, three things must be determined. From the specifications, the kind of material to be used for each caulking or sealing job must be noted; on the plans, the dimensions of the joint to be caulked or sealed must be measured, noting any requirements for backer rods and the total meters estimated. With this information, the applicable coat per linear meter can be selected from the manual and multiplied by the total length in meters to provide an estimated total cost for each kind of caulking or sealing on the job.

Insulation

Insulation is used to reduce heat transfer through the exterior enclosure of the building. The type and form of this insulation will vary according to where it is located in the structure and the accessibility of the space it occupies. Major insulation types include mineral granules, fibers and foams, vegetable fibers and solids, plus plastic foams. These materials may be required around foundations, on inside walls, or under roofing. The cost of insulation depends on the type of material, its form (loose, granular, batt, or boards), its thickness in millimeters, the method of installation, and the total area in square meters.

Shingles

Most residences and many other types of buildings have sloping roofs covered with shingles. The material used in shingles varies from the more common granular-covered asphalt and fiberglass units to wood, metal, tile, concrete, or slate.

The first steps in estimating the cost of a shingle roof are to determine the type of material specified, the size and weight of the shingles, and the method of installation. This information will permit the estimator to accurately select the installed cost of the roofing material per square meter.

Roofing and Siding

Many types of roofing and siding, in addition to shingles, are used on commercial and industrial buildings. These are made of several kinds of materials and in many forms.

The materials used in roofing and siding panels include aluminum, mineral fiber cement, epoxy, fiberglass, steel, and vinyl. Other materials used for roofing are asphalt and coal tar, asphalt and tar felt, and base sheets. Most of the latter materials are used in job-fabricated, built-up roofs and as backing for other materials such as shingles.

Single-Ply Roofs

Since the early 1970s the use of single-ply roofing membranes in the construction industry has been on the rise. Market surveys recently have shown that of all the single-ply systems being installed, about one in three is on new construction. Materially, these roofs are more expensive than other more conventional roofs; however, labor costs are much lower because of faster installation. Reroofing represents the largest market for single-ply roofs today.

Single-ply roof systems are normally installed in one of the following ways:

   a. Loose-Laid & Ballasted — Generally the easiest to install; however, some special consideration must be given where flashings are attached. The membrane is typically fused together at the seams, stretched-out flat, and then ballasted with stone (38 mm @ 49-59 kg/m^2) to prevent wind blow-off. This extra dead load must be considered during design stages. It is particularly important if reroofing over an existing built-up roof that already weighs 49-73 kg/m^2. A slip-sheet or vapor barrier is sometimes required to separate the new roof from the old.

b. Partially-Adhered — This method of installation uses a series of bar or point attachments that adhere the membrane to a substrate. The membrane manufacturer typically specifies the method to be used based on the material and substrate. Partially-adhered systems do not use ballast material. A slip-sheet may be required.

c. Fully-Adhered — Generally the most time consuming of the single-plys to install because these employ contact cement, cold adhesive, or hot bitumen to uniformly adhere the membrane to the substrate below. Only manufacturer-approved insulation board should be used to receive the membrane. No ballast is required. A slip sheet may be necessary.

The materials available can be broken into three categories:

- Thermo-Setting: EPDM, Neoprene and PIB
- Thermo-Plastic: Hypalon, PVC and CPE
- Composites: Modified Bitumen

Each category has its own unique requirements and performance characteristics. Most are available in all three installation methods.

Sheet Metal Work

Sheet metal work included in this division is limited to that used on roofs or sidewalls of buildings and is usually on the exterior and exposed to the weather. Many of the items covered are wholly or partially prefabricated with labor added for installation. Several are materials with labor added for on-site fabrication.

Pricing shop-made items such as downspouts, drip edges, expansion joints, gravel stops, gutters, reglets, and termite shields requires that the estimator determine the type of material, the size and shape of the fabricated section, and the linear meters of the item. From this data an accurate unit can be selected and multiplied by the meter to obtain a total cost.

Some roofing systems, particularly single-ply, require flashing materials unique to that system.

The cost of items like copper roofing and metal flashings is estimated in a similar manner, except that unit costs are per square meter.

Roof Accessories

Roof accessories are items found on the roof that may not be a part of the weatherproofing system but are there for another purpose. All accessories that are a standard size are priced per installed unit. These include ceiling, roof and smoke vents, and snow guards.

Skylight costs are listed per square meter with unit costs varying in steps as the nominal size of individual units increases.

Figure II.17 is the quantity takeoff for the thermal and moisture protection portion of the sample project.

COST ANALYSIS

PROJECT:	Repair Garage	CLASSIFICATION:	Division 7			SHEET NO.	8 of 15
LOCATION:		ARCHITECT:	As Shown			ESTIMATE NO:	
TAKE OFF BY:	ABC	QUANTITIES BY: ABC	PRICES BY: DEF	EXTENSIONS BY: DEF		DATE:	1993
						CHECKED BY:	GHI

DESCRIPTION	SOURCE			QUANT	UNIT	MATERIAL		LABOR		EQUIPMENT		SUBCONTRACT		TOTAL
						COST	TOTAL	COST	TOTAL	COST	TOTAL	COST	TOTAL	COST TOTAL
Moisture / Thermal														
Perimeter Insulation, 50 mm Polystyrene, 1.4 m2 · K/W	072	109	0700	61	m2	3.44	210	2.98	182					
Roof Insulation, Fiberglass 57 mm, 1.4 m2 · K/W	072	203	0800	372	m2							12.90	4799	
Coal Tar w/ Gravel Surface, 3 Plies Glass Fiber, Mopped	075	102	4800	372	m2							18.05	6715	
Al. Gravel Stop, 150 mm x 1 mm	077	105	0300	92	m							16.00	1472	
Al. Gutter, 127 mm, Enameled	076	205	0400	31	m							12.55	389	
Al. Downspout, 76 mm x 102 mm, Enameled	076	201	0400	12	m							10.80	130	
Plastic Skylight, Dbl., 1.5 m x 0.9 m	078	101	0600	8.4	m2							219.00	1840	
Division 7 Total							$210		$182				$15,344	

Note: m2 = square meter

Figure II.17

Division 8
Doors and Windows

With the ever increasing emphasis on thermal conductivity, design and use of fenestration is undergoing a dramatic change. Triple glazing, reflecting glass, absorbing glass, and no glass at all are becoming common in today's buildings. In spite of these changes in design and specifications, the estimating of this division remains unchanged. It is essentially the tabulation of the units for each opening.

Changes Required for Metric

Doors
To convert to metric, door dimensions will change from a height of 6'-8" to 2050 or 2100 mm, and from 7'-0" to 2100 mm. Door widths will change from 2'-6" to 750 mm, from 2'-8" to 800 mm, from 2'-10" to 850 mm, from 3'-0" to 900 mm, and from 3'-4" to 1000 mm. Door thicknesses will remain the same, as will door materials and hardware.

Associations on Metric
The Steel Door Institute (SDI) is addressing metrication through soft conversions, providing metric conversion charts on their new documents. The investment in re-tooling required to produce hard metric products is cited as a major obstacle for the relatively small producers of steel doors.

The National Wood Window and Door Association (NWWDA) is also anticipating soft conversions for the short-term.

Windows
Commercial window systems are available in hard metric sizes. Some of the manufacturers who produce metric sizes are listed in Appendix B.

Estimating Procedures

Metal Doors and Frames – Quantity Survey
A proper door schedule on the architectural drawings for a building will identify each opening in detail. A typical door and frame schedule is shown in Figure II.18 from *Means Forms for Building Construction Professionals*. If the architectural drawings do not provide such a schedule, one should be prepared by the estimator prior to making an estimate. Define each opening in

DOOR AND FRAME SCHEDULE

PROJECT: Office Building
LOCATION:
ARCHITECT:
OWNER:
PAGE 1 OF 1
DATE 1993
BY RMW

Qty.	DOOR NO.	SIZE W (mm)	SIZE H (mm)	SIZE T (mm)	DOOR MAT.	DOOR TYPE	GLASS	LOUVER	FRAME MAT.	FRAME TYPE	JAMB	HEAD	SILL	LAB	CON	SET NO.	KEYSIDE ROOM NO.	REMARKS
2	B01	914	2134	44	Hm.	FL.	254x254	—	Hm.	Weld	A7/6	A7/7	A7/12	B		H-1	Gar.	Epoxy Paint
	B02																	
1	101	1829	2134	—	Alum.	—	—	—	Alum.	—	A8/1	A8/2	A8/3	—		H-6	Entr.	Narrow Stile - 2 Leaf w/914mm x 1829mm Transom
2	102	914	2134	44	Hm.	FL.	—	—	Hm.	Weld	A7/6	A7/7	A7/12	—		H-2	Ext.	Insul.
	103																	N.R.P. Hinges
2	104	914	2032	44	Birch	SC/FL	—	—	Hm.	KD	A7/8	A7/9	A7/10	C		H-8	—	Painted
	105																	
4	106	914	2032	44	Birch	SC/FL	—	—	Hm.	KD	A7/8	A7/7	A7/10	—		H-4	—	Stain + Clear Finish
	107																	
	108																	
	109																	

Figure II.18

accordance with the items in the schedule and any other pertinent data. Further information should be carefully reviewed in the specifications under Division 8.

For your quantity survey, combine all like doors and frames, checking off each as you go to ensure that none are left out. Count the total number of openings, making certain that two doors have been included where double doors are used. Important details to check for are:

- Material
- Gauge
- Size
- Core Material
- Fire Rating Label
- Finish
- Style

Leave a space in the tabulation on the Quantity Sheet for casings, stops, grounds, and hardware. This can be done either on the same sheet or on separate sheets.

Wood and Plastic Doors – Quantity Survey

The Quantity Survey for Wood and Plastic Laminated Doors is identical to that for Metal Doors. Prehung doors and windows are becoming prevalent in the industry. For these, locksets and interior casings are not included. As these are usually standard for a number of doors in any particular building, they need only to be counted. Remember that exterior prehung doors need casings on the interior, and interior prehung doors need casings on both sides.

Special Doors – Quantity Survey

As indicated in *Means Building Construction Cost Data, Metric Version*, Special Doors include a full spectrum of door types from acoustical to vertical lift. Study the specifications very carefully for:

- Ratings
- Finishes
- Gauges
- Operation Mechanisms
- Materials

Doors – Pricing

Many types of doors and their installation are listed in *Means Building Construction Cost Data, Metric Version*. The installation of most doors has been figured for two carpenters. If your work rules or operating methods use an apprentice or laborer, adjust your installation cost accordingly. The estimator should also adjust the productivity in accordance with his or her records or judgment.

Entrances and Storefronts

Entrances and storefronts are almost all special designs or combinations of units to fit a unique situation. The estimator should submit the plans and specifications to an installer for takeoff and pricing. The general procedure for the installer's takeoff is as follows.

Stationary Units

- Determine height and width of each like unit.
- Determine mm of intermediate, horizontal, and vertical members rounded to next highest 10 mm.
- Determine number of joints.

Entrance Units
- Determine total mm of framing of each type required and round to next 10 mm.
- Determine number of joints.
- Determine special frame hardware per unit.
- Determine special door hardware per unit.
- Determine threshold and closers.

Pricing
- List the material and labor prices for each basic unit.
- List the unit material and labor prices for each extra horizontal and vertical member.
- List a unit labor and material price for each joint.
- Add one setup charge for each type of unit.

Add to above the cost for glass and glazing.

Metal Windows – Quantity Survey

As with doors, a window schedule should be found on the architectural drawings. If there is none, use a form or modify an existing form such as a door schedule, filling out all details found in the plans and specifications. Items to pay particular attention to are:

- Material
- Gauge
- Glazing (type of glass and setting specifications)
- Screens
- Trim
- Hardware

When using prehung units, be sure to add interior trim and stools.

Wood Windows – Quantity Survey

Takeoff for wood windows is the same as for metal windows. They should be shown on the same schedule and given the same special attention.

Windows – Pricing

After combining all like units on the Quantity Sheet, transfer these to a pricing sheet. The material price can be obtained from *Means Building Construction Cost Data, Metric Version*. Determine the unit installation costs for each type of unit and add this to the material cost. Now extend each line times the number of units and add the extensions for a total.

Finish Hardware and Specialties

If there is space on the door and window schedules for hardware, now is the time to fill it in. Remember that most prehung doors and windows do not include locksets. Some casement, awning, and jalousies include cranks and locks. Be certain to check the specifications for:

- Base Metal
- Service (Heavy, Light, Medium)
- Finish
- Any other special detail
- See also Specialties, Division 10

Check the specifications and code for panic devices and handicap devices. Public law requires accessibility for the handicapped. Certain hardware devices may be a special type as a result. Metal thresholds and astragals are also included in this division. Weatherstripping may or may not be included with prehung doors and windows. Finish hardware is difficult to price, even from a catalog, when the particular items are specified.

Prices shown in *Means Building Construction Cost Data, Metric Version* are average contractor prices if finish hardware is purchased for an entire job.

With respect to hardware, only hinges have been included in the installation cost of doors. Kick plates, panic devices, door closers and locksets, and so on, must be priced for material and labor separately.

Glass and Glazing – Quantity Survey

Glazing quantities are a function of material, method, and length to be glazed. Quantities are, therefore, measured in square meters (length + width).

Be certain to read the specifications carefully as there are many different grades and thicknesses and other variables in glass. *Means Building Construction Cost Data, Metric Version* lists many possible variables.

Pricing for glazing is usually done by mm. Prices for glass are given in *Means Building Construction Cost Data, Metric Version* by the square meter by size ranges.

The estimator should transfer quantities from the Quantity Sheet to a pricing sheet and compute costs from company records, foreman consultation, and any other means available.

Window/Curtain Walls

Average costs for window/curtain walls can be found in *Means Building Construction Cost Data, Metric Version*. With the increasing awareness of heat loss, translucent, insulated, fiberglass panels are being used for window walls. It's not a bad idea to get a price quote from a specialty installer.

Figure II.19 is the quantity takeoff for the doors and windows for the sample project.

COST ANALYSIS

PROJECT: Repair Garage	SHEET NO. 9 of 15
LOCATION:	ESTIMATE NO:
TAKE OFF BY: ABC	DATE: 1993
QUANTITIES BY: ABC	CHECKED BY: GHI
ARCHITECT:	
CLASSIFICATION: Division 8	
PRICES BY: DEF	EXTENSIONS BY: DEF

DESCRIPTION	SOURCE	QUANT	UNIT	MATERIAL COST	MATERIAL TOTAL	LABOR COST	LABOR TOTAL	EQUIPMENT COST	EQUIPMENT TOTAL	SUBCONTRACT COST	SUBCONTRACT TOTAL	TOTAL COST	TOTAL
Doors, Windows & Glass													
Door Frame, KD, 915 mm x 2135 mm, 18 Ga.	081 118 0100	7	Ea.	65.50	459	23.50	165	1.03	7				
Door, 20 Ga., 915 mm x 2135 mm 44 mm Tk	081 103 1060	7	Ea.	181.00	1267	22.00	154	0.97	7				
Ovhd. Door, 24 Ga., 3050 mm x 3050 mm	083 604 2650	9	Ea.	385.00	3465	208.00	1872						
Al. Window, Sliding, Ins., 1525 mm x 915 mm	085 204 4400	3	Ea.	165.00	495	46.00	138						
Hinges, High Frequency, 114 mm x 114 mm, USP	087 116 1000	10.5	Pair	36.50	383								
Door Closer, Reg. Arm Adj.	087 206 0020	2	Ea.	91.00	182	31.00	62						
Door Closer, Hold Open, Non-Sized	087 206 0240	5	Ea.	107.00	535	31.00	155						
Lockset, H.D., Keyed Single Cyl.	087 120 1400	7	Ea.	137.00	959	18.70	131						
Division 8 Total					$7,745		$2,676		$14				

Figure II.19

Division 9
Finishes

Interior finishes collectively represent a significant cost component in building construction. Since they cover all interior surface areas, the material quantities and variety of materials can be extensive. Estimates should, therefore, be carefully checked, to make sure that every item has been included, and to avoid minor mathematical errors that can lead to major misrepresentations of cost.

In performing quantity takeoffs for interior finishes, consideration should be given to waste allowances, particularly for materials with patterns. Furthermore, since there can be a tremendous cost range for interior finishes, the specifications should be closely reviewed to ensure that the materials in the estimate match the specified quality.

Changes Required for Metric

Lathing and Plastering
The International Institute for Lath and Plaster reports no imminent changes to the materials used for this aspect of construction, brought about by metrication. Formulas used for material mixtures are expressed in parts (ratios), which can be carried out in metric just as they would be in Imperial units. Coverages will be expressed in square meters rather than square feet.

Ceiling Systems
Grids and lay-in ceiling tile, air diffusers, and lighting fixtures will change from 2' x 2' to 600 x 600 mm, and from 2' x 4' to 600 x 1200 mm. There will be no change in grid profiles, tile thicknesses, air diffuser capacities, fluorescent tubes, or means of suspension.

Suspended ceiling systems must switch to a new hard metric size in full scale metric construction. Fortunately, many domestic manufacturers currently make these metric sizes. Some of the suspended ceiling product manufacturers who offer metric sizes are listed in Appendix B.

Raised Floor Systems
Grids and lay-in floor tile will change from 2' x 2' to 600 x 600 mm. Grid profiles, tile thicknesses, and support systems will remain the same.

Virtually all manufacturers currently offer 600 x 600 mm grid flooring. See Appendix B for product sources.

Drywall

To convert to metric, drywall width will change from 4'-0" to 1200 mm. Height will change from 8'-0" to 2400 mm, and from 10'-0" to 3000 mm. There will be no change in thickness, and fire, acoustic, and thermal ratings will remain the same.

Paint and Wallcoverings

Paint will continue, for the short term at least, to be available in Imperial measure sizes (gallons). However, conversion of units will be required to express coverage from square foot to square meter per liter, and productivity will be expressed in square meters.

Tile, Terrazzo and Flooring

The Tile Council of America (TCA) reports that most U.S. tile manufacturers are continuing to define their products in terms of the English Imperial system. When ISO standards have been completed and adopted, it is expected that tile manufacturers will produce metric-sized products.

Estimating Procedures

Lathing and Plastering

For our estimating of this section, we shall proceed in the normal erection sequence of studs, furring, lath, accessories, and so on.

Metal stud takeoff is measured on a meter basis for both stud and runner. For plastering, studs usually are spaced at 400 mm O.C.

Lath is estimated by square meters for both gypsum and metal lath, plus a usual 5% allowance for waste. Furring, channels, and accessories are measured by meters. An extra meter should be allowed for each accessory miter or stop.

The plaster is also estimated by square meter. Deductions for openings vary by preference from zero deduction on new work to 50% of all openings over 0.60 m in width. Some estimators deduct a percentage of the total for openings. One square meter extra should be allowed for each meter of horizontal interior or exterior angles located below the ceiling level. Also, double the areas of small radius work.

In pricing the work, consideration should be given to the class of the work. Basically there are two classes:

- Ordinary or commercial with waves 3 mm to 10 mm in 3 meters, angles and corners fairly true.
- First quality with variations less than 2 mm in 3 meters.

Labor for first quality work is approximately 20% more than ordinary plastering.

Each room should be measured as perimeter times maximum wall height. Ceiling areas equal length times width.

Wood plaster grounds are usually installed by carpenters but should be measured when taking off the plaster requirements.

Drywall Takeoff

In estimating accessories, studs, track, and acoustical caulking are measured by the meter. Usual stud spacing is 600 mm O.C. except where specified otherwise. Drywall taping is figured by the square meter. Gypsum wallboard comes in thicknesses of 6.35 mm to 16 mm and lengths from 1800 mm to 4200 mm. No deduction should be made for materials for door or window openings under 3 m^2. T & G coreboard can be obtained in 25 mm thicknesses for solid wall and shaft work.

Different types of partition construction should be listed separately on the quantity takeoff sheets. There may be walls with studs of various widths, including double studded and similar or dissimilar surface materials. Acoustical requirements vary and wall systems are tallied separately. Shaft work is usually of different construction from surrounding partitions requiring separate listing and pricing of the work.

Acoustical Treatment

Acoustical systems fall into several categories; e.g., sound barriers of soft dense metals or fibers of either batt or board. Ceiling tiles are either applied directly to a backer or are installed on a suspension system. Tile materials can be mineral, glass or wood fibers, or metal pans with sound-absorbing pads. The takeoff of these materials is by the square meter of area with a 5% allowance for waste.

Tile and Terrazzo

Tile and terrazzo are taken off on a square meter basis. Trim and base are measured by the meter. Accent tiles are listed per each. Two basic methods of installation are used: mud set and thin set. Mud set is about 30% more expensive than thin set. In terrazzo work, be sure to include the meterage of embedded decorative strips, grounds, machine rubbing, power cleanup, and so on.

Flooring

This section covers carpeting, composition, resilient and wood flooring, stair treads, and risers. Carpeting and pads are estimated by the square meter. Carpet is available in roll widths of 2.7 m (9'), 3.7 m (12'), and 4.6 m (15'). The most economical width should be used, as waste must be figured into the total quantity. The architect or owner may require a special laying pattern with seams in predetermined locations. This can drastically affect waste.

Resilient flooring is also measured by the square meter for all types. Base is estimated by the meter. If adhesive materials are to be quantified, these are estimated at a specified coverage rate by the liter.

Wood flooring is available in strip, parquet, or block configuration. The latter two types are usually set in adhesives with quantities estimated by the square meter. There are three basic grades of flooring: first, second, and third, plus combination grades of "second and better." There are also color grades and special grade labels for different kinds of lumber; e.g., oak, maple, pecan, beech. The estimator should be acquainted with these and the associated price differences. The laying pattern will influence labor costs. In addition to the material labor for laying wood floors, it will be necessary to make allowances for sanding and finishing these areas, unless the flooring is prefinished.

Painting

This is one area where bids vary to a greater extent than almost any other construction item. This is largely because there are many methods to measure surfaces to be painted.

Wall Coverings

Wall coverings are often estimated by the roll. Single rolls contain approximately 3.35 m^2, which forms the basis of determining the number of rolls required. However, wall coverings are usually sold in double or triple roll bolts.

The area to be covered is measured, length by height of wall above baseboards, to get the square meters of each wall. This figure is divided by 2.8 to obtain the number of single rolls, allowing 0.56 m^2 of waste per roll.

With vinyls and grass cloths requiring no pattern match, a waste allowance of 10% is normal; 0.3 m^2/roll. Wall coverings requiring a pattern match need about 25% – 30% waste; 0.8 – 1 m^2/roll. Waste can run as high as 50% – 60% on wall coverings with a large, bold, or intricate pattern repeat.

Figure II.20 is the quantity takeoff for finishes for the sample project.

COST ANALYSIS

PROJECT:	Repair Garage			CLASSIFICATION:	Division 9							SHEET NO.	10 of 15
LOCATION:				ARCHITECT:								ESTIMATE NO:	
TAKE OFF BY:	ABC			QUANTITIES BY: ABC	PRICES BY: As Shown	EXTENSIONS BY: DEF						DATE: 1993	
												CHECKED BY: GHI	

DESCRIPTION	SOURCE			QUANT	UNIT	MATERIAL		LABOR		EQUIPMENT		SUBCONTRACT		TOTAL
						COST	TOTAL	COST	TOTAL	COST	TOTAL	COST	TOTAL	TOTAL
Finishes														
C.T. Cove Base, Thin Set 108 mm x 108 mm High	093	102	0700	26	m	8.25	215	8.65	225					
C.T. Walls, Thin Set, 108 mm x 108 mm	093	102	5400	33	m2	21.50	710	19.15	632					
Acoust. Ceil. Bds. Min Fiber Al. Face, 610 mm x 1219 mm	095	104	0930	47	m2	11.20	526	3.09	145					
Class A T-Bar, Susp. 610 mm x 1219 mm	091	304	0050	47	m2	3.77	177	2.51	118					
Floor Tile, 305 mm x 305 mm x 3 mm Tk Vinyl	096	601	7600	47	m2	18.30	860	4.05	190					
Cove Base, Vinyl, 102 mm High	096	601	1150	38	m	1.71	65	1.96	74					
Subtotal							$2,553		$1,385					

Note: m2 = square meter

Figure II.20a

COST ANALYSIS

PROJECT:	Repair Garage	CLASSIFICATION:	Division 9 (cont'd)		SHEET NO.	11 of 15
LOCATION:		ARCHITECT:			ESTIMATE NO:	
TAKE OFF BY:	ABC	QUANTITIES BY: ABC	PRICES BY: As Shown	EXTENSIONS BY: DEF	DATE: 1993	CHECKED BY: GHI

DESCRIPTION	SOURCE			QUANT	UNIT	MATERIAL		LABOR		EQUIPMENT		SUBCONTRACT		TOTAL	
						COST	TOTAL	COST	TOTAL	COST	TOTAL	COST	TOTAL	COST	TOTAL
Finishes (cont'd)															
Paint Block, Smooth Finish,															
Brushwork, Primer	099	224	2100	818	m2	0.97	793	1.53	1252						
2 Coats Paint	099	224	2800	818	m2	1.40	1145	2.48	2029						
Paint Flush Door & Frame															
914 mm x 2134 mm,															
Brushwork, Primer	099	216	0500	6	Ea.	1.28	8	10.25	62						
2 Sds; Paint 2 Coats, 2 Sds	099	216	1000	12	Ea.	1.24	15	10.90	131						
Subtotal							1961		3472						
Page 10 Subtotal							2553		1385						
Page 11 Subtotal							1961		3472						
Division 9 Total							$4,514		$4,857						

Note: m2 = square meter

Figure II.20b

Division 10
Specialties

Division 10 is essentially a "count" division. Each of the items within this division must be counted separately, with the exception of partitions, which are measured in meters.

Changes Required for Metric

> Construction Specialties cover a wide range of items, from portable partitions to bath accessories. They are generally soft converted for projects that require quantities in metric, unless they are manufactured overseas, or by a U.S. manufacturer that exports a significant amount of product.

Estimating Procedures

The first step in estimating Division 10 is to read the specifications and make a list of materials and the accepted manufacturers of each item. Next, a takeoff is made from the drawings for each of these items in the list. Since most of these items are included in *Architect's First Source* or *Sweet's Catalog*, it is a good idea to get a postcard off to the manufacturer or local representative if there is sufficient time or to telephone for an up-to-date material price quotation.

The bold-face subdivision listings in *Means Building Construction Cost Data, Metric Version* serve as a handy checklist for items normally included in Division 10.

All of the items in this division should be calculated on a current delivered material price from a supplier. The labor is estimated for the complete installation of the item from initial receiving to final installation. Where the item is of sufficient size, such as a flagpole, equipment costs must also be calculated and included in the installation costs.

Figure II.21 is the quantity takeoff of specialty items for the sample project.

COST ANALYSIS

| PROJECT: | Repair Garage | CLASSIFICATION: | Division 10 | | | | SHEET NO. | 12 of 15 |

PROJECT: Repair Garage
LOCATION:
CLASSIFICATION: Division 10
ARCHITECT:
TAKE OFF BY: ABC
QUANTITIES BY: ABC
PRICES BY: As Shown
EXTENSIONS BY: DEF
CHECKED BY: GHI
ESTIMATE NO:
DATE: 1993
SHEET NO. 12 of 15

DESCRIPTION	SOURCE			QUANT	UNIT	MATERIAL		LABOR		EQUIPMENT		SUBCONTRACT		TOTAL	
						COST	TOTAL	COST	TOTAL	COST	TOTAL	COST	TOTAL	COST	TOTAL
Specialties															
Toilet Partitions, P. E. Floor Mtd., Headrail Braced	101	602	2700	1	Ea.	710.00	710	62.50	63						
Wire Partition, 1219 mm x 2438 mm	106	052	0400	10	Ea.	94.00	940	16.25	163						
Swinging Door, 914 mm x 2134 mm No Transom	106	052	2100	1	Ea.	325.00	325	62.50	63						
Mirror, SS w/ Shelf, 457 mm x 610 mm	108	204	3500	2	Ea.	67.50	135	9.35	19						
Towel Dispenser w/ Waste Receptical, Flush Mounted	108	204	0610	2	Ea.	305.00	610	18.70	37						
Toilet Tissue Dispenser, SS Surface Mtd., Double Roll	108	204	6200	2	Ea.	21.00	42	7.80	16						
Division 10 Total							$2,762		$359						

Figure II.21

Division 11
Architectural Equipment

This division generally refers to the total architectural equipment required by the type of construction; e.g., library equipment such as book stacks, newspaper racks, and desks. Often this equipment will be purchased directly by the owner and installed by the contractor. It's a good idea to add into the estimate an allowance (of about 10% of the equipment cost, if it can be determined) to cover the risk and handling costs of contractor-installed equipment. It's not unusual for the contractor to pick up these items and store them until ready for use.

Changes Required for Metric

Architectural Equipment covers a range of items, most of which are metricated using a soft conversion. As many equipment items are counted as individual units, metric dimensions are, for the most part, not an issue. In some cases, however, equipment manufactured abroad or by U.S. manufacturers who have a significant export market may be sized to metric dimensions.

Estimating Procedures

Similar to Division 10, this division involves a counting type of measurement system. A thorough reading of the specifications and direct contact with the manufacturer will produce an up-to-date material cost. The cost of installation can be easily estimated from the *Means Building Construction Cost Data, Metric Version* crew labor costs.

This is another division in which it pays to determine whether the material shown on the drawings or specified elsewhere should actually be included under this division. The bold-face listing in *Means Building Construction Cost Data, Metric Version* can serve as a checklist to ensure that items have not been omitted. Be careful to determine whether the equipment is to be supplied by the contractor or the owner and by whom it will be installed.

It is necessary to evaluate this type of equipment to determine what is required to support it, such as:

- Concrete
- Miscellaneous Iron
- Rough Carpentry
- Mechanical Coordination
- Electrical Requirements

This division includes equipment that can be packaged and delivered completely or partially factory assembled; the estimator must investigate these variables.

There are no Division 11 requirements for the sample project.

Division 12
Furnishings

Division 12 is best defined as furniture designed for specific uses, such as for dormitories, hospitals, hotels, offices and restaurants. Window treatments are also included in Division 12. Furnishings may be listed in the budget estimate to help determine the total financial investment, though they are usually not a part of the actual construction contract. If the furnishings are built-in, then additional man-hours for unpacking, installing, and clean-up are necessary and must be calculated. In most cases, architects will separately specify and arrange for the purchase and installation of furnishings outside of the contract for construction.

Changes Required for Metric

> Many types of furnishings, such as systems furniture, are used overseas in soft-converted sizes without difficulty. Since major retooling by product manufacturers would be required to produce hard metric-sized furnishings, it is expected that metrication requirements will be met with soft conversions for the foreseeable future.
>
> Systems furniture manufacturers that have product literature with metric dimensions are listed in Appendix B.

Estimating Procedures

The furnishings that are specified and/or shown on the drawings are counted as units. Each item should be counted and measured; this information should be given to manufacturers or their representatives for the latest delivered material price. The installation of these items can be calculated based on a per unit price. Often this equipment is purchased directly by the owner.

The furnishings should be given the same consideration described in Division 11.

There are no Division 12 requirements for the sample project.

Division 13
Special Construction

This division is concerned with specialized subsystems included in a building project. Most of the items in this division are constructed by specialty contractors who assemble prefabricated packages. The final cost figure is furnished by the specialty contractor after the exact requirements of the project have been specified.

Changes Required for Metric

> Special Construction items, like Equipment and Specialties, are generally metricated using a soft conversion. Many Special Construction projects, such as saunas and silos, are quantified as individual units. Some may be imported (and therefore sized in metric) or produced in the U.S. (in metric) for export.

Estimating Procedures

It is a good idea to go over this portion of the work with the bidding subcontractor to determine the exact scope of his or her work and to determine those items not covered by the quotation. If the subcontractor requires services such as excavation, unloading, or temporary services, these items must be included in the general contractor's bid.

If your company is the specialty contractor, you will have more detailed information concerning the systems. The more detailed your knowledge of a system, the easier it will be to subdivide the system into cost components. Each cost component can be further subdivided into material, labor, and equipment costs, which will fully identify the direct costs for the specialty.

There are no Division 13 requirements for the sample project.

Division 14
Conveying Systems

This division contains manufactured systems for moving people and/or materials. These systems may be standard items or custom built for the project. Be aware that long lead times and installation coordination problems may exist.

Changes Required for Metric

> Conveying Systems are generally quantified as units, and therefore, metrication is not a major issue. Soft conversions are applied to many conveying systems to calculate component dimensions, speed, and so on. Otis Elevator has announced that new product design and manufacture will be in metric. Otis does manufacture cabs and other components in hard metric sizes.

The following devices are included in this division:

- Correspondence Lifts
- Dumbwaiters
- Elevators
- Escalators
- Hoists and Cranes
- Material Handling Systems
- Moving Ramps and Walks
- Pneumatic Tube Systems

Because of the specialized construction of the above units, it is almost impossible for the general estimator to price most of this equipment, except in a very broad sense. Sometimes, package units that carry standard prices are specified. For general pricing, refer to *Means Building Construction Cost Data, Metric Version*. Quotations on specified equipment should be checked against the specifications to ascertain that all requirements have been met. The installed price of the particular system should include all required inspections, tests, and permits. The party responsible for payment should also be indicated.

There are no Division 14 requirements for the sample project.

Division 15
Mechanical

Estimates for the mechanical portion of a project should always be performed by the installing subcontractors, with their specialized experience and expertise. Nevertheless, the estimator for the general contractor often requires costs to compare to subcontract bids. In most cases, adjusted square meter costs based on previous, similar projects will suffice. A systems estimate may be used if more accuracy is needed. It is, however, esssential to understand both the work and the estimating process in order to properly interpret and analyze subcontract costs, and to ensure that all requirements are met.

Changes Required for Metric

ASHRAE on Metric

The American Society of Heating, Refrigerating and Air-Conditioning Engineers (ASHRAE) has published the *ASHRAE SI Guide*, which describes the correct use of metric terminology, symbols, and units for HVAC. The chart on the next page is from that publication and gives the nominal ISO mm size for American pipe, which shall be used to replace the U.S. inch size.

Schedule designations remain the same (e.g., Schedule 40, and type K, L, M). During transition to metric, the following paragraph and chart should be placed on mechanical cover sheets.

"All sizes are industry standard ASTM A53 pipe and ASTM B88 tube designated by their nominal millimeter (mm) diameter equivalent. See the following chart."

NOMINAL SIZE	
Inch	**mm**
1/2	15
5/8	16
3/4	20
1	25
1-1/14	32
1-1/2	40
2	50
2-1/2	65
3	80
3-1/2	90
4	100
5	125
6	150
8	200
10	250
12	300

Temperature

Degrees Celsius will be used for temperature measurements in metric.

For example, GSA requires that renovation projects of an entire building's HVAC system be converted to Celsius temperature where feasible. Renovation of only a portion of the HVAC system may remain in Fahrenheit.

All major manufacturers of HVAC control systems currently offer equipment that operates in Celsius. HVAC calculations should be done in metric units.

Air Distribution

In new construction, GSA requires that a hard metric ceiling grid be used, accompanied by hard metric lay-in diffusers and registers. Metric is required for renovations, where feasible. Many manufacturers of diffusers and registers have indicated they currently offer hard metric sizes.

Ductwork

Rectangular metal ductwork is a custom made product. Hard metric sizes are available (e.g., 300 x 600 mm). Flexible round duct is specified in soft converted sizes. Units are described in the ASHRAE SI Guide. The sheet metal designation will change from gauge to millimeters. Actual thickness will remain the same. In specifications, millimeters should be used; gauge may be added.

Air Diffusers and Grilles

Lay-in air distribution grilles and diffusers will switch to hard metric sizes for metric projects. Those that are wall-mounted or ceiling-mounted in drywall will not be required to change to new metric sizes; a small modification will be made to the overall dimension of the product, rather than a complete re-engineering.

Many companies making metric sizes for Canada and other countries simply modify their existing product. For example, the actual width of a nominal 24 x 24 inch (610 x 610 mm) diffuser is usually about 23-3/4 inches (604 mm).

To produce the same product for a normal 600 x 600 grid, each edge must be slightly shorter, or about 590 mm (23-1/4"). The manufacturers that can make the new hard metric sizes for lay-in type applications are listed in Appendix B.

HVAC Controls

All of the major manufacturers of HVAC controls currently offer products that will operate in Celsius. Some of those firms are: Johnson Controls, Barber Coleman, Robertshaw, Andover, and Honeywell. Local representatives can provide ordering information.

Mechanical Equipment

Both Carrier Corporation and Trane Corporation currently offer data on medium and large scale chillers in SI metric or customary units.

Sheet Metal Ductwork

Ductwork is generally fabricated in the locale of the project for which it is to be utilized. It is provided in any size or to any scale. Metric-dimensioned ductwork is therefore readily available at little if any additional cost.

Ductwork should be specified in rounded, nominal mm size designations (e.g., 300 x 600 mm). The firms that are able to handle any size metric order are listed in Appendix B.

Pipe and Tubing

Nominal pipe and tubing designations will change from inches to millimeters. Threads will change to metric sizes. Actual cross sections of pipe and tubing will, for the short term, remain the same. In fact, pipe and tubing sizes are already the same in many parts of the world, although their thread sizes may differ, and they are known by different names. Items such as 2" pipe (which actually has neither an inside nor an outside diameter of 2") will be simply relabeled in mm.

ASTM B88M provides standardized hard metric copper tube sizes. However, these will not be used until products have been made widely available.

When Division 15 is fully metricated, on repair and remodeling construction projects there will be a requirement for transition pieces, or special connection (reducer) pipe fittings, when metric pipe and fittings tie into existing Imperial sized piping materials.

Estimating Procedures

Preparation of the mechanical portion of the cost estimate, like most other types of construction cost estimates, occurs in four basic steps.

- General review of all plans and specifications.
- Material quantity takeoff.
- Pricing of material and labor required for installation.
- Addition of all identified costs and application of adjustment factors to arrive at the final price.

Step One. General Review

When a proposed project is of relatively large scale and complete plans and specifications are available to mechanical contractors for bidding the work, the first step in preparing an estimate is to review the plans and specifications, making notes of any special or unique requirements. Also, while it is not unusual to see something on the plans that is not reflected in the specifications, or vice versa, careful notes should be made of any contradictions, because these will require resolution before a meaningful estimate can be put together. It is also advisable to look at the other drawings for information concerning site plan, orientation, elevations, access, and so on. During this review the drawing scales should be compared. Plumbing and HVAC designs are frequently prepared by a consulting engineer not associated with the architects, which sometimes leads to inconsistencies in scale and location.

Forms are very useful in the preparation of estimates. This is especially true for the mechanical trades, which include a wide variety of items and materials. Careful measurements and a count of components cannot compensate for an oversight such as forgetting to include pipe insulation. A well-designed form acts as a checklist, a guide for standardization, and a permanent record.

Step Two. Material Takeoff

Plumbing

As in any estimate, the first step in preparing a plumbing estimate is to visualize the scope of the job by scanning all the drawings and specifications. The next step is to make a material takeoff summary sheet. Leaf through the plumbing section of the specifications, listing each item heading on a summary sheet. This will help in remembering various components as they are located on the drawings and, for major items, can serve as a quantity checkoff list. Also include labor such as cleaning, adjusting, testing, balancing, and so on, since these will not show up on the drawings.

Takeoff of Pipe and Fittings

Pipe runs for any type of system consist of straight section and fittings of various shapes, styles, and purposes. Depending on personal preference, separate forms may be used for pipe and fittings, or they may be combined on one sheet. The estimator will start at one end of each system and, using a tape, ruler, or wheeled indicator at the corresponding scale, will measure and record the straight lengths of each size of pipe. On passing a fitting, record this with a check in the appropriate column. Colored pencils may be used to mark runs that have been completed or termination points on the main line where measurements are stopped so that a branch may be taken off. Care must be taken so that any changes in material are noted.

Takeoff of Fixtures

Fixture takeoff is usually nothing more than counting the various types, sizes, and styles, and entering them on a fixture form. It is important, however, that each fixture be fully identified. A common source of error is to overlook what does or does not come with the fixture, such as faucets, tailpiece, flush valve, and so on. Equipment such as pumps, water heaters, water softeners, and all items not previously counted are listed at this time. The various counts should then be totaled and transferred to the summary sheet. Miscellaneous items that add to the plumbing contract should also be listed.

Fire Protection Systems

The takeoff of fire protection systems is very much like that of plumbing—the pipe loops are measured and fittings, valves, sprinkler heads, alarms, etc., are counted. Special requirements are noted, as well as conditions that would affect job performance and cost.

Heating, Ventilation, and Air Conditioning

Heating, ventilation, and air conditioning sheet metal (ducts, plenums, etc.) are usually estimated on a weight basis. The lengths of the various sizes are measured and recorded on a worksheet. The weight per meter of length is then determined. A count must also be made of all the duct-associated accessories such as fire dampers, diffusers, registers, and so on. This may be done during the duct takeoff; however, it is usually less confusing to make a separate count.

Items of equipment must be counted; however, it is also necessary to note sizes, capacities, controls, special characteristics, and features. Weight and size may be important if the unit is especially large or is going into comparatively close quarters.

Step Three. Material and Labor Pricing

Unit pricing can be accomplished in various ways. The most obvious is to pick up a phone and call the supplier and ask for a quote. To this quoted price, add the mechanical contractor's labor, plus overhead and profit costs. While this method is usually the most accurate, it is also the most time-consuming.

Supply of Material with Installation by Others

In this case, only the material is needed, with its overhead and profit. The item is located and its bare material cost increased by 10% for overhead and profit. This is the price times quantity that is listed on the summary sheet.

Installation of Material Supplied by Others

In this case a labor price may be located in the book next to the item to be installed. The bare labor cost must be used and then multiplied by correct quantity, plus overhead factors and profit.

Estimating Methods

Some items in Division 15 are priced by units not normally found on drawings. For example, the duct may be given on the plans and specifications by length and gauge. This has to be converted to total kilograms of material. Using geometric equations, the surface area of other shapes can be determined and this value multiplied by the weight in kilograms per square meter of the proper gauge of material. This answer is the one necessary to utilize book costs.

Figure II.22 is the quantity takeoff for mechanical items for the sample project.

COST ANALYSIS

PROJECT:	Repair Garage	CLASSIFICATION:	Division 15						SHEET NO.	13 of 15
LOCATION:		ARCHITECT:	ABC						ESTIMATE NO:	
TAKE OFF BY:	ABC	QUANTITIES BY: ABC	PRICES BY: DEF	EXTENSIONS BY: DEF					DATE: 1993	
									CHECKED BY: GHI	

DESCRIPTION	SOURCE			QUANT	UNIT	MATERIAL		LABOR		EQUIPMENT		SUBCONTRACT		TOTAL	
						COST	TOTAL	COST	TOTAL	COST	TOTAL	COST	TOTAL	COST	TOTAL
Mechanical															
Wall Hung W.C. w/ Flush Valve	152	180	3100	2	Ea.							435.00	870	*	
Rough In	152	180	3200	2	Ea.							415.00	830	*	
Wall Hung Urinal, V.C.	152	168	3000	1	Ea.							600.00	600	*	
Rough In	152	168	3300	1	Ea.							291.00	291	*	
Wall Hung Lav., V.C., 457 mm x 381 mm w/ Back Splash	152	136	6000	1	Ea.							273.00	273	*	
Rough In	152	136	6960	1	Ea.							495.00	495	*	
Lav., Vanity Top, V.C., 508 mm x 432 mm, Single Bowl	152	136	2960	1	Ea.							267.00	267	*	
Rough In	152	136	3580	1	Ea.							335.00	335	*	
Subtotal													$3,961		

Figure II.22a

COST ANALYSIS

SHEET NO. 14 of 15

PROJECT: Repair Garage
CLASSIFICATION: Division 15 (cont'd)
ESTIMATE NO:
LOCATION:
ARCHITECT:
DATE: 1993
TAKE OFF BY: ABC
QUANTITIES BY: ABC
PRICES BY: DEF
EXTENSIONS BY: DEF
CHECKED BY: GHI

DESCRIPTION	SOURCE			QUANT	UNIT	MATERIAL COST	MATERIAL TOTAL	LABOR COST	LABOR TOTAL	EQUIPMENT COST	EQUIPMENT TOTAL	SUBCONTRACT COST	SUBCONTRACT TOTAL	TOTAL COST	TOTAL
Mechanical (cont'd)															
Water Cooler, 31 L/hr, Wall	153	105	0180	1	Ea.							570.00	570	*	
Stainless Steel Cabinet	153	105	0640	1	Ea.							56.00	56	*	
Rough In	153	105	9800	1	Ea.							199.00	199	*	
Hot Water Heater, Elec. 3 kW, 45.4 L/hr	153	110	4100	1	Ea.							1500	1500	*	
Water Control Pipes & Fittings Allowance 50% of Fittings Marked *														3143	
Fire Exting. 4.3 kg, ABC	154	125	2080	8	Ea.							58.50	468		
Oil Interceptor, 1262 mL/s	152	128	4120	1	Ea.							845.00	845		
Gas Heater, Suspended 38 kW	155	480	2100	6	Ea.							890.00	5340	**	
Vents & Pipes @ 25% of Heater Allowance **														1335	
Rooftop Unit, 10 kW Cooling, 18 kW Heating	157	180	1100	1	Ea.							4375	4375		
Subtotal															17831
Page 13 Subtotal															3961
Page 14 Subtotal															17831
Division 15 Total															$21,792

Figure II.22b

Division 16
Electrical

As with Division 15, estimates for the electrical work on a project should always be performed by the installing subcontractor. Nevertheless, it is important here as well, for the general contractor estimator to have a good understanding of the requirements and costs. This can be difficult because the amount of detail shown on electrical drawings is perhaps the least of any for the major trades. As a result, the electrical estimator must not only count the items shown on the drawings, but also envision the completed design, including fittings, hangers, fasteners, devices, and cover plates.

Changes Required for Metric

Metrication will cause few complications for Division 16. The same measurement units that are used today will continue to be used; e.g., volts, amps, and watts. Estimators will, of course, need to measure the lengths of cable and conduit in lineal meters.

Metric considerations for takeoff and estimating electrical work primarily affect the designation of electrical conduit. Nominal conduit designations will change from inches to millimeters, although actual conduit cross-sections will remain the same.

Lighting Fixtures
When a hard metric 600 mm x 600 mm or 600 mm x 1200 mm grid is installed, hard metric fixture sizes must be used. A good number of firms can modify their existing fixtures to fit in the metric ceiling grid and still utilize currently used standard bulb sizes. The price of lighting fixtures will stay the same. Some firms indicate metric production capability and a willingness to bid hard metric jobs, and increase the lighting fixtures price by 10 to 15%.

NEMA on Metric

Figure II.23 shows National Electrical Manufacturers Association (NEMA)-approved designations for electrical conduit.

Electrical wire will not change at this time. Existing American Wire Gauge (AWG) sizes will remain the same.

It is important to pay close attention to the scale on drawings, making sure that sizes of conduit and wire are expressed consistently.

Estimating Procedures

Once plans and specifications for a project are complete, the following procedure for an electrical takeoff should be followed:

a. Read electrical specifications thoroughly.

b. Scan electrical plans; check other sections of specifications for effect on electrical work (in particular, mechanical and site work).

c. Check architectural and structural plans to review how electrical work should be installed.

d. Check fixture and power symbols.

e. Clarify with the architects or engineers any gray areas in the scope of work. Addenda may be necessary to explain certain items of work so that responsibility for the task is defined.

f. Immediately contact suppliers, manufacturers, and subsystem specialty contractors to get their quotations, sub-bids, and drawings.

NEMA Approved Metric Size Designations for EMT, IMC, GRC (Electrical Conduit)			
Trade Size	**Metric Size Designations**		
1/2"	16 EMT	16 IMC	16 GRC
3/4"	21 EMT	21 IMC	21 GRC
1"	27 EMT	27 IMC	27 GRC
1-1/4"	35 EMT	35 IMC	35 GRC
1-1/2"	41 EMT	41 IMC	41 GRC
2"	53 EMT	53 IMC	53 GRC
2-1/2"	63 EMT	63 IMC	63 GRC
3"	78 EMT	78 IMC	78 GRC
3-1/2"	91 EMT	91 IMC	91 GRC
4"	103 EMT	103 IMC	103 GRC
5"			129 GRC
6"			155 GRC
For products produced to standards detailed in NEMA bulletin 5RN 2189			

Based on IEC (International Electrotechnical Commission) designations

Figure II.23

Quantity Takeoff

1. Fixture takeoff is usually the easiest way to become familiar with the plans. Try to take off one bay or section at a time. Mark each with colored pencils and list as you proceed, carefully measuring lengths to fixtures. Check the site plans for exterior fixtures.
2. Next, take off switches, receptacles, telephone, fire alarm, sound and television outlets, plus any other system outlets.
3. While making a takeoff of the above work, identify the items or systems listed below and determine the amount of takeoff necessary, as opposed to the materials that will be included in a quotation or sub-bid.

   - Switchgear and Switchboards
   - Transformers
   - Motors and Controls
   - Motor Starters
   - Service Panels
   - Electric Heating
   - Line Voltage Controls
   - Safety Switches
   - Hazardous Locations
   - Bus Duct
   - Generators

4. Feeder conduit and wire should be carefully taken off using a scale, not a rotometer. This is more accurate. Large conduit and wire are expensive and require extensive labor. Switchboard location should be marked on each floor using column lines; this makes horizontal runs easier to take off. Distance between floors should be marked on riser plans and added to horizontal for a complete feeder run. Conduits should be measured at right angles unless it has been determined that they can be installed in floors. Elbows, terminations, bends, or expansion joints should be taken off at this time and marked in proper columns on feeder takeoff sheets.
5. Branch circuits should be taken off using a rotometer. Care should be taken to start and stop accurately on outlets. Start with two-wire circuits and mark with colored pencil as you take off. Calculate stub ups or drops for wiring devices. It is easy to take your switch count and multiply by distance from switch to ceiling.

Receptacles are handled similarly, depending on whether a run is in the ceiling or floor. In many cases there are two conduits acceptable to a receptacle box as you feed in and out to the next outlet.

The methods just described apply to most concrete and fireproof construction. In some metal stud partitions it is acceptable to go horizontally from receptacle to receptacle. Horizontal runs are usually made in wood partitions. In hung ceilings, the specifications should be checked to see if straight runs are allowed.

Transfer of Material Takeoff to Price Sheets

Use a separate price sheet for the following categories:

- Switchgear
  - Transformers
  - Panels
  - Bus Duct
- Motors Safety
  - Switches
  - Starters
  - Controls
  - Supports
- Sound System including Boxes, Conduit, & Wire
- TV System including Boxes, Conduit, & Wire
- Other Systems including Boxes, Conduit, & Wire
- Feeders
  - Junction Boxes
- Branch Circuits
- Wiring Devices and Outlet Boxes
- Lighting Fixtures
  - Lamps
  - Outlet Boxes
- Fire Alarm including Boxes, Conduit, and Wire
- Service
- Miscellaneous

By keeping systems on separate sheets, it is easy to isolate various costs. This helps when the owner wants to know where to cut costs. It also provides a breakdown that can be submitted to the general contractor when billing.

1. Switchboards, panels, and transformers can normally be entered directly onto pricing sheets as the takeoff is made. Some panels and transformers require a steel frame support from floor to ceiling; these should be added at this time.
2. Feeders should be totaled by size and transferred to price sheets. If you use standard feeder takeoff sheets, the wire column will reflect longer lengths than conduit because of the added wire used in panels and switchboard to make connections. This added length can be determined by checking a manufacturer's catalog, or at times they are shown on plans. Conduit, cable supports, and expansion joints should be totaled and transferred at this time.
3. Branch circuits should be transferred to price sheets. Add about 5% to conduit quantities for normal waste. On wire, add 10% to 12% waste to make connections. Conduit fittings such as locknuts, bushings, couplings, expansion joints, and fasteners should be added. Two conduit terminations per box are average.
4. Wiring devices should be entered with plates, boxes, and plaster rings if necessary.
5. Motors, safety switches, starters, controls, and supports should have junction boxes, supports, wiring troughs, and wire connectors, including tape or connector covers. Add short runs of Greenfield, Sealtite, connectors, and wire for motor connections.
6. Fixtures should have outlet boxes, plaster rings, Greenfield with connectors, and fixture wire if needed. Fixture wire should be included for connections in long runs of fluorescent fixtures. Include fixture supports as needed and wire nuts or other connectors.
7. Other systems should be handled in the same way, adding miscellaneous boxes, etc.
8. Service should be taken off from plans. Be careful to add all miscellaneous fittings not shown but needed to make a complete system.

Pricing the Estimate

Material

Prices are available from electrical supply houses, pricing services, or cost data books. Cost data books are easy to use and include labor and markup costs, as well as material costs. Special item costs must come from supply houses or manufacturers.

Labor

Labor units are readily available in estimating books and often are calculated in hours or fractions of hours (0.50 = 30 minutes). These units must be carefully used, taking job conditions into account.

Estimate Summary

Transfer all individual pricing sheet totals to a recapitulation or summary sheet. This sheet is used also as a checklist and should consider these questions.

- Should you add hours for excessive handling of materials? Will the foreman be working or supervising? Is premium time required? All of these items depend upon job conditions and must be considered.
- The price of bond, insurance, and all taxes such as Social Security, sales, unemployment insurance, and workers' compensation must be included, as well as permit fees and utility charges.
- Is an office and storage trailer required? Who furnishes temporary light and power? Trucking material to job site?
- Is the prime contractor reliable and does he pay promptly?
- What is the coffee break policy? Portal to portal policy?
- Is there electric in sections of other trades?
- Have there been addenda?
- Should bids go to all prime contractors?

Hoisting, cutting, patching, painting access panels — when whatever is necessary has been added for these items, total columns.

Multiply labor hours by average labor rate. This should include the actual amount paid per man plus any amount paid to insurance funds or to the union in worker benefits per hour. Enter job expenses and total with labor and material to get the prime cost.

Now add overhead. This consists of the cost of running the business in general: rent, light, heat, taxes, office payroll, or any cost not covered on the job site.

Now add the profit, which is the fee the owner feels is needed to do the job. This is the amount the company needs to maintain a healthy growing business. Consideration should be given to the amount of financial risk involved and whether the present workload and bonding capacity would enable the company to do the job without a manpower or financial strain.

The takeoff form for the electrical portion of the sample project appears in Figure II.24.

COST ANALYSIS

PROJECT: Repair Garage
LOCATION:
TAKE OFF BY: ABC
CLASSIFICATION: Division 16
ARCHITECT:
QUANTITIES BY: ABC **PRICES BY:** As Shown
EXTENSIONS BY: DEF
SHEET NO. 15 of 15
ESTIMATE NO:
DATE: 1993
CHECKED BY: GHI

DESCRIPTION	SOURCE			QUANT	UNIT	MATERIAL COST	MATERIAL TOTAL	LABOR COST	LABOR TOTAL	EQUIPMENT COST	EQUIPMENT TOTAL	SUBCONTRACT COST	SUBCONTRACT TOTAL	TOTAL COST	TOTAL TOTAL
Electrical															
Panel 120/208V Main CB, 3PH 4W, 225A, 32 Cirs.	163	245	2200	1	Ea.							2225	2225		
Rigid Galv. Stl. Cond. 50 mm	160	205	1870	9.15	m							38.00	348		
#2/0 THW Wire Stranded	161	165	0280	32	m							7.80	250		
#6 THWN Wire Stranded	161	165	1350	11	m							2.56	28		
Receptacle Duplex	162	320	2460	30	Ea.							10.10	303		
Receptacle GFI, EMT & Wire	168	170	4380	8	Ea.							120.00	960		
Wall Plate 1 Gang	162	320	2600	52	Ea.							5.95	309		
Steel Outlet Box, Square	162	110	0150	52	Ea.							17.65	918		
Stl Outlet Box, Plaster Rings	162	110	0300	52	Ea.							5.95	309		
Toggle Switch, Single Pole	162	320	2600	14	Ea.							5.95	83		
Fluor. Strip Fixture 1200 mm, 40 Watts	166	130	2300	45	Ea.							71.50	3218		
Emergency Lights, Lead Battery	166	110	0500	6	Ea.							355.00	2130		
Safety Switches, 3P, 30A	163	360	2910	1	Ea.							182.00	182		
19 kW HVAC Motor Connection	160	275	0200	1	Ea.							132.00	132		
Combination, with Motor Circuit Protector, 4 kW, Size 0	163	130	0700	1	Ea.							740.00	740		
Smoke Detector	168	120	5200	3	Ea.							113.00	339		
Break Glass Station	168	120	7000	3	Ea.							87.50	263		
Fire Alarm Horn	168	120	5800	1	Ea.							83.00	83		
EMT Conduit, 15 mm	160	205	5000	625	m							7.55	4719		
#14 THWN Solid	161	165	0920	375	m							0.97	364		
#12 THWN Solid	161	165	0940	1312	m							1.18	1548		
#10 THWN Solid	161	165	0960	296	m							1.41	417		
Division 16 Total													$19,867		

Figure II.24

Part III
METRIC IN DESIGN

liter
Ω kg mm
joule
tonne pascal Celsius
m³

Part III
Metric in Design

Part III addresses metric considerations in drawings and specifications. It includes issues such as preferred metric scales and specs, use of metric terminology in specifications and drawings, and slope ratios. A set of metric architectural drawings shows the guidelines in practice.

Drawings

According to the Construction Metrication Council of the National Institute of Building Sciences, construction drawings will change in the following ways to meet metric requirements:

- Units of linear measure will change to millimeters (from feet and inches) for all building dimensions. On large site plans and civil engineering drawings, meters will be used, carried to one, two, or three decimal places.
- Drawing scales will change from inch fractions-to-feet (e.g., 1/2" = 1'-0") to true ratios (e.g., 1:20). (See Figure III.1, a chart that compares metric to customary scales and ratios.)
- The International Standards Organization (ISO) drawing sizes are preferred metric sizes for design drawings. They are "A" Series.
    - A0   1189 x 841 mm, 46.8 x 33.1 inches
    - A1   841 x 594 mm, 33.1 x 23.4 inches
    - A2   594 x 420 mm, 23.4 x 16.5 inches
    - A3   420 x 297 mm, 16.5 x 11.7 inches
    - A4   297 x 210 mm, 11.7 x 8.3 inches

A0 is the base drawing size, with an area of one square meter. Smaller sizes are obtained by halving the long dimension of the previous size. All A0 sizes have a height-to-width ratio of 1 to the square root of 2. Metric drawings can, of course, be made on any size paper.

Slopes

Slopes are typically expressed in ratios when using metric dimensionally. Figure III.2 compares slope ratios with the customarily used angles (pitches) and percentages.

A Word of Caution

Dual units (both inch-pound and metric) should never be used on the same drawings. This practice increases the time required for dimensioning, doubles the chance of errors, and contributes confusion to the drawings. Centimeters are not used in construction drawings.

Figures III.3a through III.3g are drawings from a set of plans for a commercial building in metric dimensions.

Specifications

For metric projects, the units of measure used in specifications are millimeters (for feet-inches), square meters (for square feet), and cubic meters (for cubic yards). Liters are used for fluid volumes.

Like drawings, specifications should not include dual units, unless an inch-pound measure is used to clarify an otherwise unfamiliar metric measure. In such cases, the inch-pound unit should be placed in parentheses after the metric term; for example, 7460 W or 7.46 kW (10 horsepower). Unit conversions should be checked carefully to ensure that rounding does not exceed allowable tolerances.

Metric vs. Customary Scales and Ratios		
Metric Scales	**Customary Ratio**	**Customary Scales**
1:5	1:4	3" = 1'0"
1:10	1:18	1 1/2" = 1'0"
	1:12	1" = 1'0"
1:20	1:16	3/4" = 1'0"
	1:24	1/2" = 1'0"
1:50	1:48	1/4" = 1'0"
1:100	1:96	1/8" = 1'0"
1:200	1:92	1/16" = 1'0"
1:500	1:384	1/32" = 1'0"
	1:480	1" = 40'0"
	1:600	1" = 50'0"
1:1000	1:960	1" = 80'0"
	1:1200	1" = 100'0"
1:2000	1:2400	1" = 200'0"
1:5000	1:4800	1" = 400'0"
	1:6000	1" = 500'0"
1:10 000	1:10 560	6" = 1 mi
	1:12 000	1" = 1000'0"
1:25 000	1:21 120	3" = 1 mi
	1:24 000	1" = 2000'0"
1:50 000	1:63 360	1" = 1 mi
1:100 000	1:126 720	1/2" = 1 mi

Figure III.1

| Expression of Slope ||||
Ratio Y/X	Angle	Angle (Rad)	Percentage (%)
Shallow slopes			
1:100	0° 34'	0.010 0	1
1:67	0° 52'	0.015 0	1.5
1:57	1°	0.017 5	1.75
1:50	1° 09'	0.020 0	2
1:40	1° 26'	0.025 0	2.5
1:33	1° 43'	0.030 0	3
1:29	2°	0.034 9	3.5
1:25	2° 17'	0.039 9	4
1:20	2° 52'	0.049 9	5
1:19	3°	0.052 4	5.25
Slight slopes			
1:17	3° 26'	0.059 9	6
1:15	3° 48'	0.066 4	6.7
1:14.3	4°	0.069 8	7
1:12	4° 46'	0.083 2	8.3
1:11.4	5°	0.087 3	8.75
1:10	5° 43'	0.099 8	10
1:9.5	6°	0.104 7	10.5
1:8	7° 07'	0.124 5	12.5
1:7.1	8°	0.139 6	14
1:6.7	8° 32'	0.149 0	15
1:6	9° 28'	0.165 2	16.7
1:5.7	10°	0.174 5	17.6
1:5	11° 19'	0.197 5	20
1:4.5	12° 30'	0.218 2	22.2
1:4	14° 02'	0.245 0	25
Medium slopes			
1:3.7	15°	0.261 8	25.8
1:3.3	16° 42'	0.291 5	30
1:3	18° 26'	0.321 7	33.3
1:2.75	20°	0.349 1	36.4
1:2.5	21° 48'	0.380 5	40
1:2.4	22° 30'	0.392 7	41.4
1:2.15	25°	0.436 3	46.6
1:2	26° 34'	0.453 7	50
1:1.73	30°	0.532 6	57.5
1:1.67	30° 58'	0.540 5	60
1:1.5	33° 42'	0.588 0	67
1:1.33	36° 52'	0.643 4	75
1:1.2	40°	0.698 1	84
1:1	45°	0.785 4	100
Steep slopes			
1.19:1	50°	0.872 7	119
1.43:1	55°	0.959 9	143
1.5:1	56° 19'	0.982 7	150
1.73:1	60°	1.047 2	173
2:1	63° 26'	1.107 1	200
2.15:1	65°	1.134 5	215
2.5:1	68° 12'	1.190 3	250
2.75:1	70°	1.221 7	275
3:1	71° 34'	1.249 1	300
3.73:1	75°	1.309 0	373
4:1	75° 58'	1.325 3	400
5:1	78° 42'	1.373 5	500
5.67:1	80°	1.396 3	567
6:1	80° 32'	1.405 6	600
11.43:1	85°	1.483 5	1143
∞	90°	1.570 8	∞

Figure III.2

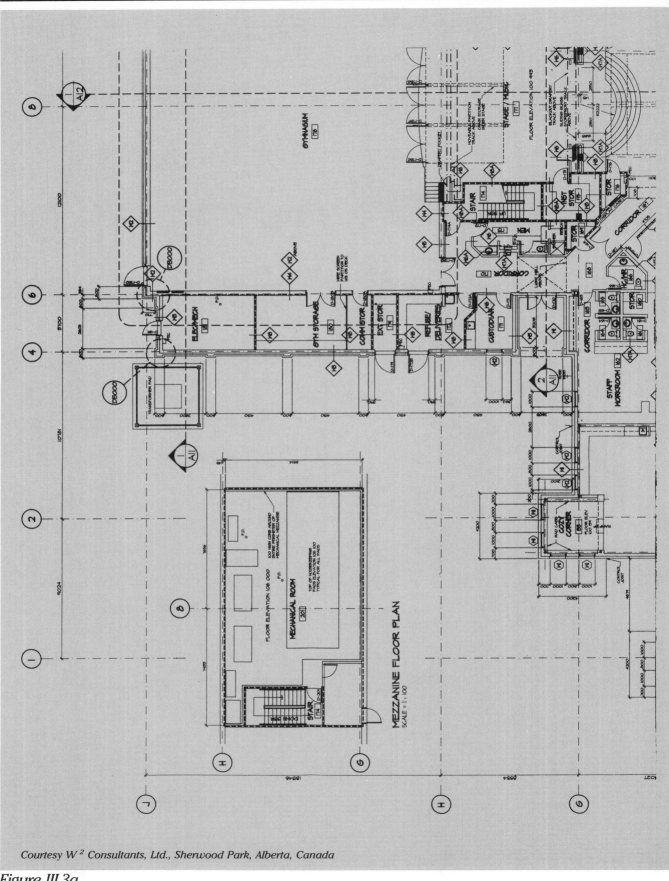

Courtesy W² Consultants, Ltd., Sherwood Park, Alberta, Canada

Figure III.3a

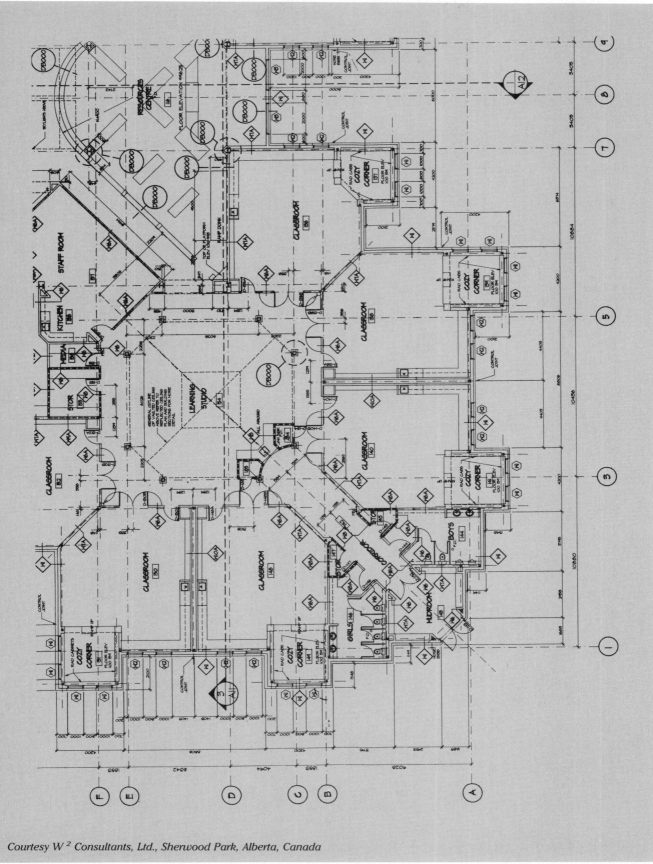

Courtesy W² Consultants, Ltd., Sherwood Park, Alberta, Canada

Figure III.3b

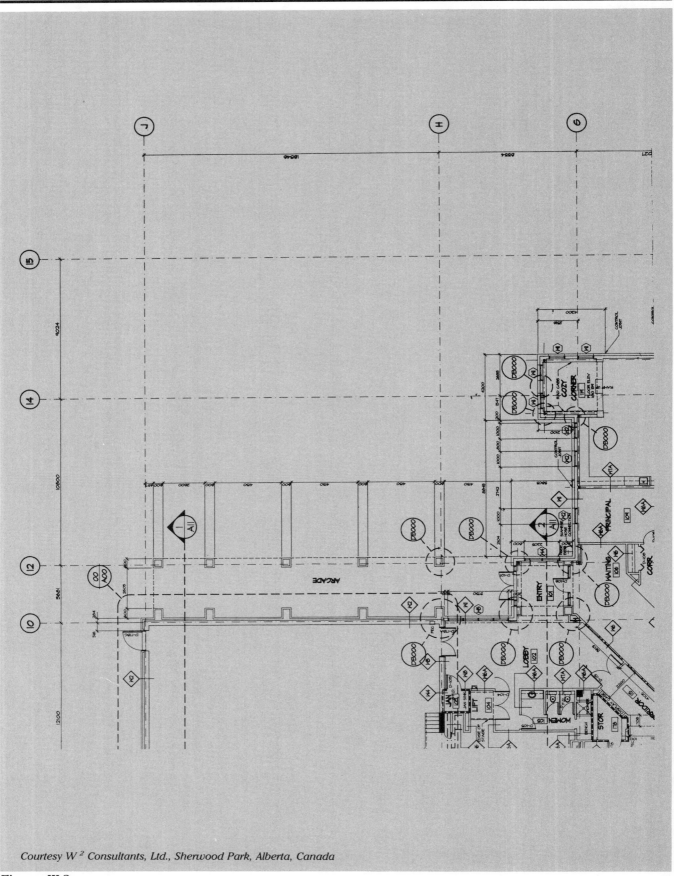

Courtesy W² Consultants, Ltd., Sherwood Park, Alberta, Canada

Figure III.3c

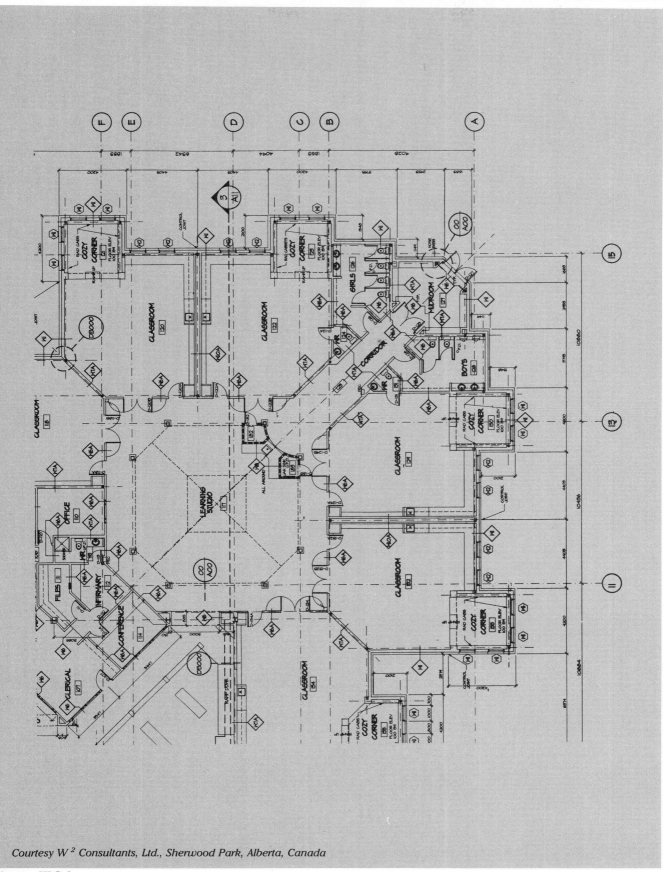

Courtesy W² Consultants, Ltd., Sherwood Park, Alberta, Canada

Figure III.3d

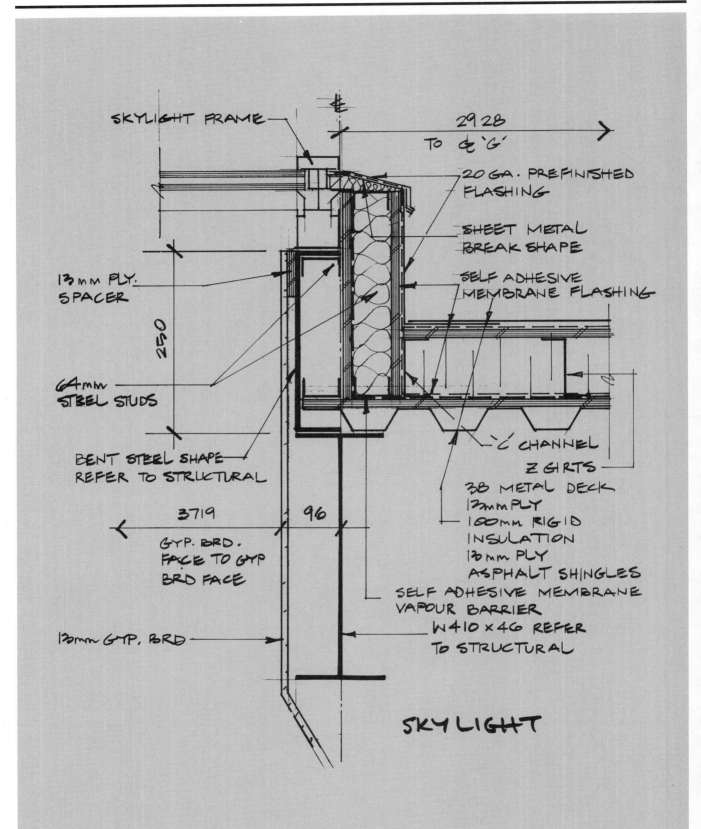

Courtesy W² Consultants, Ltd., Sherwood Park, Alberta, Canada

Figure III.3e

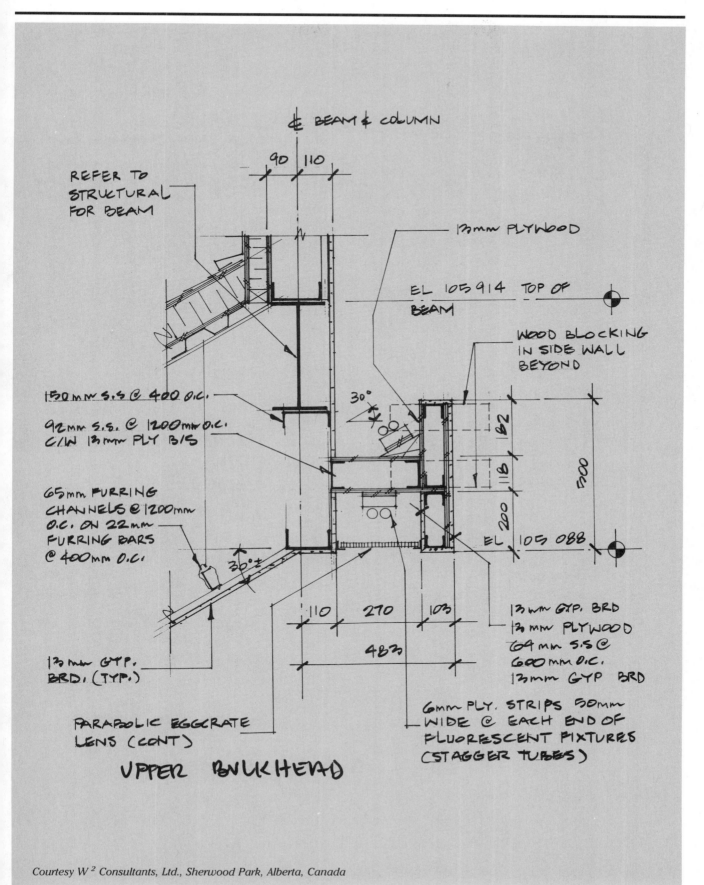

Courtesy W² Consultants, Ltd., Sherwood Park, Alberta, Canada

Figure III.3f

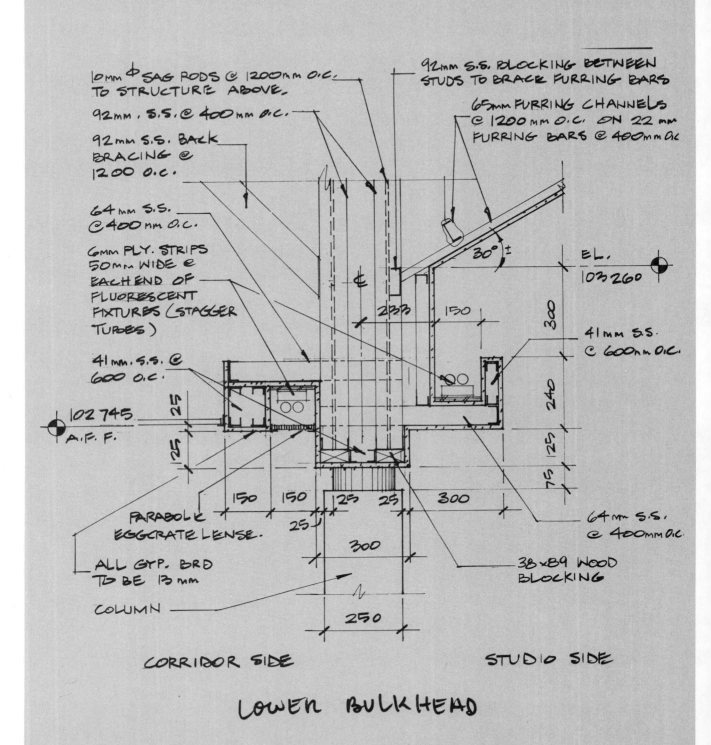

Courtesy W² Consultants, Ltd., Sherwood Park, Alberta, Canada

Figure III.3g

Appendix

Appendix A: Metric Conversion Tables   145
Appendix B: Metric Product Sources   155
Appendix C: Metric References   163
Appendix D: Professional Associations   169
Glossary of Terms   177
Index   179

Appendix A
Metric Conversion Tables

Appendix A

Standard and Metric Linear and Area Conversion Tables

Linear Conversions

Inches	Feet	Yards	Rods	Miles	Centi-meters	Meters	Kilo-meters
1	0.083	0.028	0.005	—	2.540	0.0254	—
12	1	0.333	0.061	0.0002	30.480	0.305	0.0003
36	3	1	0.182	0.0006	91.440	0.914	0.0009
0.3937	0.033	0.011	—	—	1	0.01	—
39.37	3.281	1.094	0.199	0.0006	100	1	0.001

Inches	Feet	Yards	Rods	Furlongs	Centi-meters	Meters	Kilo-meters
198	16.5	5.5	1	0.003	0.025	5.029	0.005
	5280	1760	320	1	8	1 609 347	1.609
	660	220	40	0.125	1	201.168	0.201
	3280.83	1093.61	198.838	0.621	4.971	1000	1

Area Conversions

Square Inches	Square Feet	Square Yards	Acres	Square Centi-meters	Square Meters	Hectares	Square Kilo-meters
1	0.007	—	—	6.452	0.0006	—	—
144	1	0.111	0.00002	929.034	0.093	—	—
1296	9	1	0.0002	8361.31	0.836	—	—
0.155	0.001	—	—	1	0.0001	—	—
1549.997	10.764	1.196	0.0002	10 000	1	0.0001	—

Square Inches	Square Feet	Square Yards	Acres	Square Miles	Square Meters	Hectares	Square Kilo-meters
	43 560	4840	1	0.002	4046.87	0.405	0.004
	27 878 400	3 097 600	640	1	2 589 998	258.999	2.590
	107 638.7	11 959.9	2.471	0.004	10 000	1	0.01
	10 763 867	1 195 985	247.104	0.386	1 000 000	100	1

Basic Metric Units and Prefixes

Basic Metric Units

Quantity	Unit
length	meter (m)
mass	kilogram (kg)
time	second(s)
electric current	ampere (A)
temperature (thermodynamic)	kelvin (K)
amount of substance	mole (mol)
luminous intensity	candela (cd)

Prefixes for Metric Units

Multiple and Submultiple		Prefix	Symbol
1 000 000 000 000	$= 10^{12}$	tera	T
1 000 000 000	$= 10^{9}$	giga	G
1 000 000	$= 10^{6}$	mega	M
1000	$= 10^{3}$	kilo	k
100	$= 10^{2}$	hecto	h
10	$= 10$	deka	da
0.1	$= 10^{-1}$	deci	d
0.01	$= 10^{-2}$	centi	c
0.001	$= 10^{-3}$	milli	m
0.000 001	$= 10^{-6}$	micro	μ
0.000 000 001	$= 10^{-9}$	nano	n
0.000 000 000 001	$= 10^{-12}$	pico	p
0.000 000 000 000 001	$= 10^{-15}$	femto	f
0.000 000 000 000 000 001	$= 10^{-18}$	atto	a

Appendix A

Metric Conversion Table

	Inches to Centimeters	Feet to Meters	Pounds to Kilograms		Inches to Centimeters	Feet to Meters	Pounds to Kilograms
1	2.54	0.304 8	0.453 6	51	129.54	15.544 8	23.133 6
2	5.08	0.609 6	0.907 2	52	132.08	15.849 6	23.587 2
3	7.62	0.914 4	1.360 8	53	134.62	16.154 4	24.040 8
4	10.16	1.219 2	1.814 4	54	137.16	16.459 2	24.494 4
5	12.7	1.524	2.268	55	139.7	16.764	24.948
6	15.24	1.828 8	2.721 6	56	142.24	17.068 8	25.401 6
7	17.78	2.133 6	3.175 2	57	144.78	17.373 6	25.855 2
8	20.32	2.438 4	3.628 8	58	147.32	17.678 4	26.308 8
9	22.86	2.743 2	4.082 4	59	149.86	17.983 2	26.762 4
10	25.4	3.048	4.536	60	152.4	18.288	27.216
11	27.94	3.352 8	4.989 6	61	154.94	18.592 8	27.669 6
12	30.48	3.657 6	5.443 2	62	157.48	18.897 6	28.123 2
13	33.02	3.962 4	5.896 8	63	160.02	19.202 4	28.576 8
14	35.56	4.267 2	6.350 4	64	162.56	19.507 2	29.030 4
15	38.1	4.572	6.804	65	165.1	19.812	29.488
16	40.64	4.876 8	7.257 6	66	167.64	20.116 8	29.937 6
17	43.18	5.181 6	7.711 2	67	170.18	20.421 6	30.391 2
18	45.72	5.486 4	8.164 8	68	172.72	20.726 4	30.844 8
19	48.26	5.791 2	8.618 4	69	175.26	21.031 2	31.298 4
20	50.8	6.096	9.072	70	177.8	21.336	31.752
21	53.34	6.400 8	9.525 6	71	180.34	21.640 8	32.205 6
22	55.88	6.705 6	9.979 2	72	182.88	21.945 6	32.659 2
23	58.42	7.010 4	10.432 8	73	185.42	22.250 4	33.112 8
24	60.96	7.315 2	10.886 4	74	187.96	22.555 2	33.566 4
25	63.5	7.62	11.34	75	190.5	22.86	34.02
26	66.04	7.924 8	11.793 6	76	193.04	23.164 8	34.473 6
27	68.58	8.229 6	12.247 2	77	195.58	23.469 6	34.927 2
28	71.12	8.534 4	12.700 8	78	198.12	23.774 4	35.380 8
29	73.66	8.839 2	13.154 4	79	200.66	24.079 2	35.834 4
30	76.2	9.144	13.608	80	203.2	24.384	36.288
31	78.74	9.448 8	14.061 6	81	205.74	24.688 8	36.741 6
32	81.28	9.753 6	14.515 2	82	208.28	24.993 6	37.195 2
33	83.82	10.058 4	14.968 0	83	210.82	25.298 4	37.648 8
34	86.36	10.363 2	15.422 4	84	213.36	25.603 2	38.102 4
35	88.9	10.668	15.876	85	215.9	25.908	38.556
36	91.44	10.972 8	16.329 6	86	218.44	26.212 8	39.009 6
37	93.98	11.277 6	16.783 2	87	220.98	26.517 6	39.463 2
38	96.52	11.582 4	17.236 8	88	223.52	26.822 4	39.916 8
39	99.06	11.887 2	17.690 4	89	226.06	27.127 2	40.370 4
40	101.6	12.192	18.144	90	228.6	27.432	40.824
41	104.14	12.496 8	18.597 6	91	231.14	27.736 8	41.277 6
42	106.68	12.801 6	19.051 2	92	233.68	28.041 6	41.731 2
43	109.22	13.106 4	19.504 8	93	236.22	28.346 4	42.184 8
44	111.76	13.411 2	19.958 4	94	238.76	28.651 2	42.638 4
45	114.3	13.716	20.412	95	241.3	28.956	43.092
46	116.84	14.020 8	20.865 6	96	243.84	29.260 8	43.545 6
47	119.38	14.325 6	21.319 2	97	246.38	29.565 6	43.999 2
48	121.92	14.630 4	21.772 8	98	248.92	29.870 4	44.452 8
49	124.46	14.935 2	22.226 4	99	252.46	30.175 2	44.906 4
50	127	15.24	22.68	100	254	30.48	45.36

Conversion Formulas: Inches x 2.54 = Centimeters
Feet x 0.3048 = Meters
Pounds x 0.4536 = Kilograms

Appendix A

To convert from	to	Multiply by
abampere	ampere (A)	1.000 000*E+01
abcoulomb	coulomb (C)	1.000 000*E+01
abfarad	farad (F)	1.000 000*E+09
abhenry	henry (H)	1.000 000*E−09
abmho	siemens (S)	1.000 000*E+09
abohm	ohm (Ω)	1.000 000*E−09
abvolt	volt (V)	1.000 000*E−08
acre foot[13]	cubic metre (m^3)	1.233 489 E+03
acre[13]	square metre (m^2)	4.046 873 E+03
ampere hour	coulomb (C)	3.600 000*E+03
angstrom	metre (m)	1.000 000*E−10
are	square metre (m^2)	1.000 000*E+02
astronomical unit	metre (m)	1.495 979 E+11
atmosphere, standard	pascal (Pa)	1.013 250*E+05
atmosphere, technical (= 1 kgf/cm^2)	pascal (Pa)	9.806 650*E+04
bar	pascal (Pa)	1.000 000*E+05
barn	square metre (m^2)	1.000 000*E−28
barrel (for petroleum, 42 gal)	cubic metre (m^3)	1.589 873 E−01
board foot	cubic metre (m^3)	2.359 737 E−03
British thermal unit (International Table)[14]	joule (J)	1.055 056 E+03
British thermal unit (mean)	joule (J)	1.055 87 E+03
British thermal unit (thermochemical)	joule (J)	1.054 350 E+03
British thermal unit (39°F)	joule (J)	1.059 67 E+03
British thermal unit (59°F)	joule (J)	1.054 80 E+03
British thermal unit (60°F)	joule (J)	1.054 68 E+03
Btu (International Table)·ft/(h·ft^2·°F) (thermal conductivity)	watt per metre kelvin [W/(m·K)]	1.730 735 E+00
Btu (thermochemical)·ft/(h·ft^2·°F) (thermal conductivity)	watt per metre kelvin [W/(m·K)]	1.729 577 E+00
Btu (International Table)·in/(h·ft^2·°F) (thermal conductivity)	watt per metre kelvin [W/(m·K)]	1.442 279 E−01
Btu (thermochemical)·in/(h·ft^2·°F) (thermal conductivity)	watt per metre kelvin [W/(m·K)]	1.441 314 E−01
Btu (International Table)·in/(s·ft^2·°F) (thermal conductivity)	watt per metre kelvin [W/(m·K)]	5.192 204 E+02
Btu (thermochemical)·in/(s·ft^2·°F) (thermal conductivity)	watt per metre kelvin [W/(m·K)]	5.188 732 E+02
Btu (International Table)/h	watt (W)	2.930 711 E−01
Btu (International Table)/s	watt (W)	1.055 056 E+03
Btu (thermochemical)/h	watt (W)	2.928 751 E−01
Btu (thermochemical)/min	watt (W)	1.757 250 E+01
Btu (thermochemical)/s	watt (W)	1.054 350 E+03
Btu (International Table)/ft^2	joule per square metre (J/m^2)	1.135 653 E+04
Btu (thermochemical)/ft^2	joule per square metre (J/m^2)	1.134 893 E+04

---

[13] The U.S. Metric Law of 1866 gave the relationship, 1 metre equals 39.37 inches. Since 1893 the U.S. yard has been derived from the metre. In 1959 a refinement was made in the definition of the yard to bring the U.S. yard and the yard used in other countries into agreement. The U.S. yard was changed from 3600/3937 m to 0.9144 m exactly. The new length is shorter by exactly two parts in a million.

At the same time it was decided that any data in feet derived from and published as a result of geodetic surveys within the U.S. would remain with the old standard (1 ft = 1200/3937 m) until further decision. This foot is named the U.S. survey foot.

All conversion factors for units of land measure in these tables referenced to this footnote are based on the U.S. survey foot and the following relationships: 1 fathom = 6 feet; 1 rod (pole or perch) = 16½ feet; 1 chain = 66 feet; 1 mile (U.S. statute) = 5280 feet.

[14] The Fifth International Conference on the Properties of Steam in 1956 defined the calorie (International Table) as 4.1868 J. Therefore, the exact conversion for Btu (International Table) is 1.055 055 852 62 E+03 J.

Copyright ASTM. Reprinted with permission.

Appendix A

To convert from	to	Multiply by
Btu (International Table)/(ft²·s)	watt per square metre (W/m²)	1.135 653 E+04
Btu (International Table)/(ft²·h)	watt per square metre (W/m²)	3.154 591 E+00
Btu (thermochemical)/(ft²·h)	watt per square metre (W/m²)	3.152 481 E+00
Btu (thermochemical)/(ft²·min)	watt per square metre (W/m²)	1.891 489 E+02
Btu (thermochemical)/(ft²·s)	watt per square metre (W/m²)	1.134 893 E+04
Btu (thermochemical)/(in²·s)	watt per square metre (W/m²)	1.634 246 E+06
Btu (International Table)/(h·ft²·°F) (thermal conductance)[15]	watt per square metre kelvin [W/(m²·K)]	5.678 263 E+00
Btu (thermochemical)/(h·ft²·°F) (thermal conductance)[15]	watt per square metre kelvin [W/(m²·K)]	5.674 466 E+00
Btu (International Table)/(s·ft²·°F)	watt per square metre kelvin [W/(m²·K)]	2.044 175 E+04
Btu (thermochemical)/(s·ft²·°F)	watt per square metre kelvin [W/(m²·K)]	2.042 808 E+04
Btu (International Table)/lb	joule per kilogram (J/kg)	2.326 000*E+03
Btu (thermochemical)/lb	joule per kilogram (J/kg)	2.324 444 E+03
Btu (International Table)/(lb·°F) (heat capacity)	joule per kilogram kelvin [J/(kg·K)]	4.186 800*E+03
Btu (thermochemical)/(lb·°F) (heat capacity)	joule per kilogram kelvin [J/(kg·K)]	4.184 000*E+03
Btu (International Table)/ft³	joule per cubic metre (J/m³)	3.725 895 E+04
Btu (thermochemical)/ft³	joule per cubic metre (J/m³)	3.723 402 E+04
bushel (U.S.)	cubic metre (m³)	3.523 907 E−02
calorie (International Table)[14]	joule (J)	4.186 800*E+00
calorie (mean)	joule (J)	4.190 02 E+00
calorie (thermochemical)	joule (J)	4.184 000*E+00
calorie (15°C)	joule (J)	4.185 80 E+00
calorie (20°C)	joule (J)	4.181 90 E+00
calorie (kilogram, International Table)	joule (J)	4.186 800*E+03
calorie (kilogram, mean)	joule (J)	4.190 02 E+03
calorie (kilogram, thermochemical)	joule (J)	4.184 000*E+03
cal (thermochemical)/cm²	joule per square metre (J/m²)	4.184 000*E+04
cal (International Table)/g	joule per kilogram (J/kg)	4.186 800*E+03
cal (thermochemical)/g	joule per kilogram (J/kg)	4.184 000*E+03
cal (International Table)/(g·°C)	joule per kilogram kelvin [J/(kg·K)]	4.186 800*E+03
cal (thermochemical)/(g·°C)	joule per kilogram kelvin [J/(kg·K)]	4.184 000*E+03
cal (thermochemical)/min	watt (W)	6.973 333 E−02
cal (thermochemical)/s	watt (W)	4.184 000*E+00
cal (thermochemical)/(cm²·s)	watt per square metre (W/m²)	4.184 000*E+04
cal (thermochemical)/(cm²·min)	watt per square metre (W/m²)	6.973 333 E+02
cal (thermochemical)/(cm²·s)	watt per square metre (W/m²)	4.184 000*E+04
cal (thermochemical)/(cm·s·°C)	watt per metre kelvin [W/(m·K)]	4.184 000*E+02
cd/in²	candela per square metre (cd/m²)	1.550 003 E+03
carat (metric)	kilogram (kg)	2.000 000*E−04
centimetre of mercury (0°C)	pascal (Pa)	1.333 22 E+03
centimetre of water (4°C)	pascal (Pa)	9.806 38 E+01
centipoise (dynamic viscosity)	pascal second (Pa·s)	1.000 000*E−03
centistokes (kinematic viscosity)	square metre per second (m²/s)	1.000 000*E−06
chain[13]	metre (m)	2.011 684 E+01
circular mil	square metre (m²)	5.067 075 E−10
clo	kelvin square metre per watt (K·m²/W)	1.55 E−01
cup	cubic metre (m³)	2.365 882 E−04
curie	becquerel (Bq)	3.700 000*E+10
darcy[16]	square metre (m²)	9.869 233 E−13
day	second (s)	8.640 000*E+04
day (sidereal)	second (s)	8.616 409 E+04
degree (angle)	radian (rad)	1.745 329 E−02
degree Celsius	kelvin (K)	$T_K = t_{°C} + 273.15$
degree centigrade	[see 3.4.2]	
degree Fahrenheit	degree Celsius (°C)	$t_{°C} = (t_{°F} - 32)/1.8$

---

[15] In ISO 31 this quantity is called *coefficient of heat transfer*.
[16] The darcy is a unit for measuring permeability of porous solids.

Copyright ASTM. Reprinted with permission.

Appendix A

To convert from	to	Multiply by
degree Fahrenheit	kelvin (K)	$T_K = (t_F + 459.67)/1.8$
degree Rankine	kelvin (K)	$T_K = T_R/1.8$
°F·h·ft²/Btu (International Table) (thermal resistance)[17]	kelvin square metre per watt (K·m²/W)	1.761 102 E−01
°F·h·ft²/Btu (thermochemical) (thermal resistance)[17]	kelvin square metre per watt (K·m²/W)	1.762 280 E−01
°F·h·ft²/[Btu (International Table)·in] (thermal resistivity)	kelvin metre per watt (K·m/W)	6.933 471 E+00
°F·h·ft²/[Btu (thermochemical)·in] (thermal resistivity)	kelvin metre per watt (K·m/W)	6.938 113 E+00
denier	kilogram per metre (kg/m)	1.111 111 E−07
dyne	newton (N)	1.000 000*E−05
dyne·cm	newton metre (N·m)	1.000 000*E−07
dyne/cm²	pascal (Pa)	1.000 000*E−01
electronvolt	joule (J)	1.602 19 E−19
EMU of capacitance	farad (F)	1.000 000*E+09
EMU of current	ampere (A)	1.000 000*E+01
EMU of electric potential	volt (V)	1.000 000*E−08
EMU of inductance	henry (H)	1.000 000*E−09
EMU of resistance	ohm (Ω)	1.000 000*E−09
ESU of capacitance	farad (F)	1.112 650 E−12
ESU of current	ampere (A)	3.335 6 E−10
ESU of electric potential	volt (V)	2.997 9 E+02
ESU of inductance	henry (H)	8.987 554 E+11
ESU of resistance	ohm (Ω)	8.987 554 E+11
erg	joule (J)	1.000 000*E−07
erg/(cm²·s)	watt per square metre (W/m²)	1.000 000*E−03
erg/s	watt (W)	1.000 000*E−07
faraday (based on carbon-12)	coulomb (C)	9.648 70 E+04
faraday (chemical)	coulomb (C)	9.649 57 E+04
faraday (physical)	coulomb (C)	9.652 19 E+04
fathom[13]	metre (m)	1.828 804 E+00
fermi (femtometre)	metre (m)	1.000 000*E−15
fluid ounce (U.S.)	cubic metre (m³)	2.957 353 E−05
foot	metre (m)	3.048 000*E−01
foot (U.S. survey)[13]	metre (m)	3.048 006 E−01
foot of water (39.2°F)	pascal (Pa)	2.988 98 E+03
ft²	square metre (m²)	9.290 304*E−02
ft²/h (thermal diffusivity)	square metre per second (m²/s)	2.580 640*E−05
ft²/s	square metre per second (m²/s)	9.290 304*E−02
ft³ (volume; section modulus)	cubic metre (m³)	2.831 685 E−02
ft³/min	cubic metre per second (m³/s)	4.719 474 E−04
ft³/s	cubic metre per second (m³/s)	2.831 685 E−02
ft⁴ (second moment of area)[18]	metre to the fourth power (m⁴)	8.630 975 E−03
ft/h	metre per second (m/s)	8.466 667 E−05
ft/min	metre per second (m/s)	5.080 000*E−03
ft/s	metre per second (m/s)	3.048 000*E−01
ft/s²	metre per second squared (m/s²)	3.048 000*E−01
footcandle	lux (lx)	1.076 391 E+01
footlambert	candela per square metre (cd/m²)	3.426 259 E+00
ft·lbf	joule (J)	1.355 818 E+00
ft·lbf/h	watt (W)	3.766 161 E−04
ft·lbf/min	watt (W)	2.259 697 E−02
ft·lbf/s	watt (W)	1.355 818 E+00
ft-poundal	joule (J)	4.214 011 E−02
g, standard free fall	metre per second squared (m/s²)	9.806 650*E+00

[17] In ISO 31 this quantity is called *thermal insulance* and the quantity *thermal resistance* has the unit K/W.
[18] This is sometimes called the moment of section or area moment of inertia of a plane section about a specified axis.

Copyright ASTM. Reprinted with permission.

Appendix A

To convert from	to	Multiply by
gal	metre per second squared (m/s²)	1.000 000*E−02
gallon (Canadian liquid)	cubic metre (m³)	4.546 090 E−03
gallon (U.K. liquid)	cubic metre (m³)	4.546 092 E−03
gallon (U.S. dry)	cubic metre (m³)	4.404 884 E−03
gallon (U.S. liquid)	cubic metre (m³)	3.785 412 E−03
gallon (U.S. liquid) per day	cubic metre per second (m³/s)	4.381 264 E−08
gallon (U.S. liquid) per minute	cubic metre per second (m³/s)	6.309 020 E−05
gallon (U.S. liquid) per hp·h (SFC, specific fuel consumption)	cubic metre per joule (m³/J)	1.410 089 E−09
gamma	tesla (T)	1.000 000*E−09
gauss	tesla (T)	1.000 000*E−04
gilbert	ampere (A)	7.957 747 E−01
gill (U.K.)	cubic metre (m³)	1.420 653 E−04
gill (U.S.)	cubic metre (m³)	1.182 941 E−04
grade	degree (angular)	9.000 000*E−01
grade	radian (rad)	1.570 796 E−02
grain	kilogram (kg)	6.479 891*E−05
grain/gal (U.S. liquid)	kilogram per cubic metre (kg/m³)	1.711 806 E−02
gram	kilogram (kg)	1.000 000*E−03
g/cm³	kilogram per cubic metre (kg/m³)	1.000 000*E+03
gf/cm²	pascal (Pa)	9.806 650*E+01
hectare	square metre (m²)	1.000 000*E+04
horsepower (550 ft·lbf/s)	watt (W)	7.456 999 E+02
horsepower (boiler)	watt (W)	9.809 50 E+03
horsepower (electric)	watt (W)	7.460 000*E+02
horsepower (metric)	watt (W)	7.354 99 E+02
horsepower (water)	watt (W)	7.460 43 E+02
horsepower (U.K.)	watt (W)	7.457 0 E+02
hour	second(s)	3.600 000*E+03
hour (sidereal)	second (s)	3.590 170 E+03
hundredweight (long)	kilogram (kg)	5.080 235 E+01
hundredweight (short)	kilogram (kg)	4.535 924 E+01
inch	metre (m)	2.540 000*E−02
inch of mercury (32°F)	pascal (Pa)	3.386 38 E+03
inch of mercury (60°F)	pascal (Pa)	3.376 85 E+03
inch of water (39.2°F)	pascal (Pa)	2.490 82 E+02
inch of water (60°F)	pascal (Pa)	2.488 4 E+02
in²	square metre (m²)	6.451 600*E−04
in³ (volume)[19]	cubic metre (m³)	1.638 706 E−05
in³ (section modulus)[19]	metre cubed (m³)	1.638 706 E−05
in³/min	cubic metre per second (m³/s)	2.731 177 E−07
in⁴ (second moment of area)[18]	metre to the fourth power (m⁴)	4.162 314 E−07
in/s	metre per second (m/s)	2.540 000*E−02
in/s²	metre per second squared (m/s²)	2.540 000*E−02
kayser	1 per metre (1/m)	1.000 000*E+02
kelvin	degree Celsius (°C)	$t_{°C} = T_K - 273.15$
kilocalorie (International Table)	joule (J)	4.186 800*E+03
kilocalorie (mean)	joule (J)	4.190 02 E+03
kilocalorie (thermochemical)	joule (J)	4.184 000 E+03
kilocalorie (thermochemical)/min	watt (W)	6.973 333 E+01
kilocalorie (thermochemical)/s	watt (W)	4.184 000 E+03
kilogram-force (kgf)	newton (N)	9.806 650*E+00
kgf·m	newton metre (N·m)	9.806 650*E+00
kgf·s²/m (mass)	kilogram (kg)	9.806 650*E+00
kgf/cm²	pascal (Pa)	9.806 650*E+04
kgf/m²	pascal (Pa)	9.806 650*E+00
kgf/mm²	pascal (Pa)	9.806 650*E+06
km/h	metre per second (m/s)	2.777 778 E−01

[19] The exact conversion factor is 1.638 706 4*E−05.

Copyright ASTM. Reprinted with permission.

Appendix A

To convert from	to	Multiply by
kilopond (1 kp = 1 kgf)	newton (N)	9.806 650*E+00
kW·h	joule (J)	3.600 000*E+06
kip (1000 lbf)	newton (N)	4.448 222 E+03
kip/in² (ksi)	pascal (Pa)	6.894 757 E+06
knot (international)	metre per second (m/s)	5.144 444 E−01
lambert	candela per square metre (cd/m²)	1/π *E+04
lambert	candela per square metre (cd/m²)	3.183 099 E+03
langley	joule per square metre (J/m²)	4.184 000*E+04
light year	metre (m)	9.460 55 E+15
litre[20]	cubic metre (m³)	1.000 000*E−03
lm/ft²	lumen per square metre (lm/m²)	1.076 391 E+01
maxwell	weber (Wb)	1.000 000*E−08
mho	siemens (S)	1.000 000*E+00
microinch	metre (m)	2.540 000*E−08
micron (deprecated term, use micrometre)	metre (m)	1.000 000*E−06
mil	metre (m)	2.540 000*E−05
mile (international)	metre (m)	1.609 344*E+03
mile (U.S. statute)[13]	metre (m)	1.609 347 E+03
mile (international nautical)	metre (m)	1.852 000*E+03
mile (U.S. nautical)	metre (m)	1.852 000*E+03
mi² (international)	square metre (m²)	2.589 988 E+06
mi² (U. S. statute)[13]	square metre (m²)	2.589 998 E+06
mi/h (international)	metre per second (m/s)	4.470 400*E−01
mi/h (international)	kilometre per hour (km/h)	1.609 344*E+00
mi/min (international)	metre per second (m/s)	2.682 240*E+01
mi/s (international)	metre per second (m/s)	1.609 344*E+03
millibar	pascal (Pa)	1.000 000*E+02
millimetre of mercury (0°C)	pascal (Pa)	1.333 22 E+02
minute (angle)	radian (rad)	2.908 882 E−04
minute	second (s)	6.000 000*E+01
minute (sidereal)	second (s)	5.983 617 E+01
oersted	ampere per metre (A/m)	7.957 747 E+01
ohm centimetre	ohm meter (Ω·m)	1.000 000*E−02
ohm circular-mil per foot	ohm metre (Ω·m)	1.662 426 E−09
ounce (avoirdupois)	kilogram (kg)	2.834 952 E−02
ounce (troy or apothecary)	kilogram (kg)	3.110 348 E−02
ounce (U.K. fluid)	cubic metre (m³)	2.841 306 E−05
ounce (U.S. fluid)	cubic metre (m³)	2.957 353 E−05
ounce-force	newton (N)	2.780 139 E−01
ozf·in	newton metre (N·m)	7.061 552 E−03
oz (avoirdupois)/gal (U.K. liquid)	kilogram per cubic metre (kg/m³)	6.236 023 E+00
oz (avoirdupois)/gal (U.S. liquid)	kilogram per cubic metre (kg/m³)	7.489 152 E+00
oz (avoirdupois)/in³	kilogram per cubic metre (kg/m³)	1.729 994 E+03
oz (avoirdupois)/ft²	kilogram per square metre (kg/m²)	3.051 517 E−01
oz (avoirdupois)/yd²	kilogram per square metre (kg/m²)	3.390 575 E−02
parsec	metre (m)	3.085 678 E+16
peck (U.S.)	cubic metre (m³)	8.809 768 E−03
pennyweight	kilogram (kg)	1.555 174 E−03
perm (0°C)	kilogram per pascal second square metre [kg/(Pa·s·m²)]	5.721 35 E−11
perm (23°C)	kilogram per pascal second square metre [kg/(Pa·s·m²)]	5.745 25 E−11
perm·in (0°C)	kilogram per pascal second metre [kg/(Pa·s·m)]	1.453 22 E−12
perm·in (23°C)	kilogram per pascal second metre [kg/(Pa·s·m)]	1.459 29 E−12
phot	lumen per square metre (lm/m²)	1.000 000*E+04

[20] In 1964 the General Conference on Weights and Measures reestablished the name litre as a special name for the cubic decimetre. Between 1901 and 1964 the litre was slightly larger (1.000 0.28 dm³); in the use of high-accuracy volume data of that time interval, this fact must be kept in mind.

Copyright ASTM. Reprinted with permission.

Appendix A

To convert from	to	Multiply by
pica (printer's)	metre (m)	4.217 518 E−03
pint (U.S. dry)	cubic metre (m³)	5.506 105 E−04
pint (U.S. liquid)	cubic metre (m³)	4.731 765 E−04
point (printer's)	metre (m)	3.514 598*E−04
poise (absolute viscosity)	pascal second (Pa·s)	1.000 000*E−01
pound (lb avoirdupois)[21]	kilogram (kg)	4.535 924 E−01
pound (troy or apothecary)	kilogram (kg)	3.732 417 E−01
lb·ft² (moment of inertia)	kilogram square metre (kg·m²)	4.214 011 E−02
lb·in² (moment of inertia)	kilogram square metre (kg·m²)	2.926 397 E−04
lb/ft·h	pascal second (Pa·s)	4.133 789 E−04
lb/ft·s	pascal second (Pa·s)	1.488 164 E+00
lb/ft²	kilogram per square metre (kg/m²)	4.882 428 E+00
lb/ft³	kilogram per cubic metre (kg/m³)	1.601 846 E+01
lb/gal (U.K. liquid)	kilogram per cubic metre (kg/m³)	9.977 637 E+01
lb/gal (U.S. liquid)	kilogram per cubic metre (kg/m³)	1.198 264 E+02
lb/h	kilogram per second (kg/s)	1.259 979 E−04
lb/hp·h (SFC, specific fuel consumption)	kilogram per joule (kg/J)	1.689 659 E−07
lb/in³	kilogram per cubic metre (kg/m³)	2.767 990 E+04
lb/min	kilogram per second (kg/s)	7.559 873 E−03
lb/s	kilogram per second (kg/s)	4.535 924 E−01
lb/yd³	kilogram per cubic metre (kg/m³)	5.932 764 E−01
poundal	newton (N)	1.382 550 E−01
poundal/ft²	pascal (Pa)	1.488 164 E+00
poundal·s/ft²	pascal second (Pa·s)	1.488 164 E+00
pound-force (lbf)[22]	newton (N)	4.448 222 E+00
lbf·ft	newton metre (N·m)	1.355 818 E+00
lbf·ft/in	newton metre per metre (N·m/m)	5.337 866 E+01
lbf·in	newton metre (N·m)	1.129 848 E−01
lbf·in/in	newton metre per metre (N·m/m)	4.448 222 E+00
lbf·s/ft²	pascal second (Pa·s)	4.788 026 E+01
lbf·s/in²	pascal second (Pa·s)	6.894 757 E+03
lbf/ft	newton per metre (N/m)	1.459 390 E+01
lbf/ft²	pascal (Pa)	4.788 026 E+01
lbf/in	newton per metre (N/m)	1.751 268 E+02
lbf/in² (psi)	pascal (Pa)	6.894 757 E+03
lbf/lb (thrust/weight [mass] ratio)	newton per kilogram (N/kg)	9.806 650 E+00
quart (U.S. dry)	cubic metre (m³)	1.101 221 E−03
quart (U.S. liquid)	cubic metre (m³)	9.463 529 E−04
rad (absorbed dose)	gray (Gy)	1.000 000*E−02
rem (dose equivalent)	sievert (Sv)	1.000 000*E−02
rhe	1 per pascal second [1/(Pa·s)]	1.000 000*E+01
rod[13]	metre (m)	5.029 210 E+00
roentgen	coulomb per kilogram (C/kg)	2.58 000*E−04
rpm (r/min)	radian per second (rad/s)	1.047 198 E−01
second (angle)	radian (rad)	4.848 137 E−06
second (sidereal)	second (s)	9.972 696 E−01
shake	second (s)	1.000 000*E−08
slug	kilogram (kg)	1.459 390 E+01
slug/ft·s	pascal second (Pa·s)	4.788 026 E+01
slug/ft³	kilogram per cubic metre (kg/m³)	5.153 788 E+02
statampere	ampere (A)	3.335 640 E−10
statcoulomb	coulomb (C)	3.335 640 E−10
statfarad	farad (F)	1.112 650 E−12
stathenry	henry (H)	8.987 554 E+11
statmho	siemens (S)	1.112 650 E−12
statohm	ohm (Ω)	8.987 554 E+11
statvolt	volt (V)	2.997 925 E+02

[21] The exact conversion factor is 4.535 923 7*E−01.
[22] The exact conversion factor is 4.448 221 615 260 5*E+00.

Copyright ASTM. Reprinted with permission.

Appendix A

To convert from	to	Multiply by
stere	cubic metre (m³)	1.000 000*E+00
stilb	candela per square metre (cd/m²)	1.000 000*E+04
stokes (kinematic viscosity)	square metre per second (m²/s)	1.000 000*E−04
tablespoon	cubic metre (m³)	1.478 676 E−05
teaspoon	cubic metre (m³)	4.928 922 E−06
tex	kilogram per metre (kg/m)	1.000 000*E−06
therm (European Community)[23]	joule (J)	1.055 06 E+08
therm (U.S.)[23]	joule (J)	1.054 804*E+08
ton (assay)	kilogram (kg)	2.916 667 E−02
ton (long, 2240 lb)	kilogram (kg)	1.016 047 E+03
ton (metric)	kilogram (kg)	1.000 000*E+03
ton (nuclear equivalent of TNT)	joule (J)	4.184 E+09[24]
ton of refrigeration (= 12 000 Btu/h)	watt (W)	3.517 E+03
ton (register)	cubic metre (m³)	2.831 685 E+00
ton (short, 2000 lb)	kilogram (kg)	9.071 847 E+02
ton (long)/yd³	kilogram per cubic metre (kg/m³)	1.328 939 E+03
ton (short)/yd³	kilogram per cubic metre (kg/m³)	1.186 553 E+03
ton (short)/h	kilogram per second (kg/s)	2.519 958 E−01
ton-force (2000 lbf)	newton (N)	8.896 443 E+03
tonne	kilogram (kg)	1.000 000*E+03
torr (mmHg, 0°C)	pascal (Pa)	1.333 22 E+02
unit pole	weber (Wb)	1.256 637 E−07
W·h	joule (J)	3.600 000*E+03
W·s	joule (J)	1.000 000*E+00
W/cm²	watt per square metre (W/m²)	1.000 000*E+04
W/in²	watt per square metre (W/m²)	1.550 003 E+03
yard	metre (m)	9.144 000*E−01
yd²	square metre (m²)	8.361 274 E−01
yd³	cubic metre (m³)	7.645 549 E−01
yd³/min	cubic metre per second (m³/s)	1.274 258 E−02
year (365 days)	second (s)	3.153 600*E+07
year (sidereal)	second (s)	3.155 815 E+07
year (tropical)	second (s)	3.155 693 E+07

[23] The therm (European Community) is legally defined in the Council of the European Communities Directive 80/181/EC of December 20, 1979. The therm (U.S.) is legally defined in the *Federal Register*, Vol 33, No. 146, p. 10756, of July 27, 1968. Although the European therm, which is based on the International Table Btu, is frequently used by engineers in the U.S., the therm (U.S.) is the legal unit used by the U.S. natural gas industry.

[24] Defined (not measured) value.

Copyright ASTM. Reprinted with permission.

Appendix B
Metric Product Sources

This appendix contains a listing of manufacturers who provide building products in metric dimensions or classifications. The list is based, in part, on research conducted by the General Services Administration. It is not intended to be all-inclusive, but may be useful as a starting point for those seeking to familiarize themselves with the metric estimating process.

Air Diffusers & Grilles

The most sophisticated manufacturers build their systems to a 3 mm tolerance, and there is no price change. The rest of these manufacturers shear down their larger products and charge for a custom fit job.

Waste will be a factor only until the sheet product producers start to manufacture to the 600 mm grid system.

Acutherm, Emeryville, CA, a manufacturer of VAV air distribution devices, can manufacture its products in hard metric sizes. Contact: Jim Kline, 510-428-1064.

Aireguide, Hialeah, FL, a large manufacturer of air distribution products, can make 80-90% of its products in hard metric sizes. Contact: Daryl Gray, 305-888-1631.

Carnes, Verona, WI, one of the larger manufacturers of air distribution products, regularly makes hard metric sizes. No cost premium. Contact: Dick Laughlin, 608-845-6411.

Donco Air Products, Albion, IA, a small fixture manufacturer but a major manufacturer of light troffer diffusers, can manufacture light troffer, slot, and lay-in diffusers in hard metric sizes up to 1500 mm length. Lead time is 4 weeks. No minimum order required. Contact 1: Ron Jansen, Engineering, 515-488-2211. Contact 2: Marc Vandegrift, Engineering, 515-488-2211.

Duralast, New Orleans, LA, can make its primary diffuser product in a 600 x 600 mm variation. Contact: Ron Vinson (distributor), 504-837-2346.

J & J Register, El Paso, TX, can make hard metric sizes. Contact: Chris Smith, 915-852-9111.

Juniper Industries, Middle Valley, NY, can produce metric size diffusers and grilles. Contact: Steve Liebermann, 718-326-2456.

Krueger, Inc., Tuscon, AZ, a large manufacturer of grilles and diffusers, has the capability to manufacture hard metric sizes. Contact: Steve Bowser, 602-622-7601.

Reliable Metal Products, Geneva, AL, a subsidiary of Hart & Cooley, is a medium-size manufacturer of air distribution products, and can make about 90% of its products in hard metric sizes. Contact: John Bowers, 205-684-3621.

Appendix B

Rock Island Register, Rock Island, IL, can make its standard product, a 2' x 2' diffuser, in a 600 x 600 mm size. Contact: John Howarth, 309-788-5611.

Sommerville Metalcraft, Cranfordsville, IL, can produce grilles and diffusers in hard metric sizes. Contact: Paul Moehling, 800-654-3124.

Thermo Kinetics, Greenville, SC, can make its standard grilles and diffusers in hard metric sizes. Contact: Terry Rutledge, 803-277-8080.

Titus Products, Richardson, TX, a major manufacturer of air distribution grilles and products, indicates a number of products currently available in hard metric sizes. Contact: Dave Loren, 214-699-1030.

Chillers

The larger companies have already started to convert; smaller manufacturers are able to do soft conversions to fit into metric system.

Carrier Corp. currently offers data on its large scale chillers in SI metric or customary units.

Trane Corp. currently offers data on its medium-scale chillers in SI metric or customary units.

Curtainwall Systems

Curtainwall construction varies so much from job to job that the industry is very capable of work in millimeters. It often works with millimeters. Most of the marble and stone is cut outside the U.S.

Curtainwall systems are obtainable in hard metric sizes.

Profile Systems, Gerald, MO, a subsidiary of the Maune Company, is able to manufacture curtainwall systems to any hard metric size. Contact: Grant Maune, 800-962-8100.

Kalwell Corp., Manchester, NH, recognized worldwide for its leadership role in the industry, is able to produce any size metric curtainwall system. Contact: Bruce Keller, 800-258-9777.

Kawneer Company, Norcross, GA, a major international curtainwall manufacturer, has been supplying curtainwall systems in metric units to the overseas market for years, and is fully able to handle any metric order. Contact 1: Enrique Morales, Int'l Sales Mgr., 703-433-2711, Contact 2: Edward Bugg, Asst. Engrg. Mgr., 703-433-2711.

Custom Products

The following products are custom made as a matter of course and can, therefore, be specified in any metric size without extra cost.

Glass

Precast concrete facades are custom made products, able to be made in hard metric sizes.

Rectangular metal ductwork in commercial building construction is a custom product, able to be manufactured in hard metric sizes.

Doors & Frames

The metal door industry will probably come around more slowly than most of the rest. The factors are wall thickness, drywall, metal studs, block and brick, and thickness of the door frame itself. Most manufacturers call any change from standard size (3' x 7') a custom door. Soft conversion is being done. Hard conversion will come only after the rest of the industry has converted in both residential and commercial.

Domestic manufacturers, as a matter of practice, produce hollow metal doors and wooden doors in any length and width desired. Hard metric sizes can, therefore, be specified.

Appendix B

American Steel Products, Farmingdale, NY, can make any size metric door. Contact: Hank, 516-293-7100.

Amweld Building Products, Inc., 1500 Amweld Dr., Industrial Park, Garrettsville, OH, produces hard metric doors. Call 216-527-4385.

Ceco Door Division, Oak Brook, IL, a major manufacturer in the door industry, can make any hard metric size door. Contact: Norb Bruzan, 312-242-2000.

Curries Company, Mason City, IA, produces hard metric doors. Call 515-423-1334.

Duolock, Portland, OR, a division of Alumax, a major U.S. manufacturer of aluminum products, can make any size metric frame. Contact: Clem Grant, 800-678-0566.

Fenestra Corp., Erie, PA, produces hard metric doors. Call 814-838-2001.

Kewanee Corp., Kewanee, IL, produces hard metric doors. Call 309-853-4481.

Hollow Metal Manufacturers Association, Chicago, IL, produces hard metric doors. Contact: Ed Estes, 804-583-3367.

Republic Builders Products, McKenzie, TN, produces hard metric doors. Call 901-352-3383.

Steelcraft, Cincinnati, OH, produces hard metric doors. Call 513-745-6400.

Steel Door Institute, Cleveland, OH, produces hard metric doors. Call 216-899-0010.

Timely Industries, Pacoima, CA, produces hard metric doors. Call 213-875-0124.

Drywall

The length and thickness of drywall can be changed immediately. The width is controlled by the width of the covering of the board and waste in the initial change. The largest drywall manufacturers either actively sell metric size drywall or have the capability to produce it.

Standard metric drywall width is 1200 mm. Standard stud spacing is 400 mm. Lengths are available in any size. Widths are 12.7 mm and 15.9 mm, which correspond to currently available thicknesses.

Firms offering and interested in bidding metric drywall are listed below:

USG Interiors International, Chicago, IL, Contact: Bruce Swenson, Export Sales Manager, 312-606-5831.

GoldBond, Contact: Kurt Withrock, Marketing Mgr., Gypsum Products, 704-365-7475.

Domtar, Contact: Jim Hanser, 313-930-4700.

Temple Inland, Contact: Jim Rush, 800-231-6060.

Elevators

Hard metric size is now available. There does not appear to be a cost consideration.

Otis Elevator, Farmington, CT, one of the world's largest manufacturers, announced several years ago that new product design and manufacture would be in metric to support the global market. Otis is able to provide cabs and other components in hard metric sizes. Contact: Anthony Cooney, Mechanical Engineering Manager, 203-678-2000.

Appendix B

Lighting Fixtures

Several major manufacturers of fluorescent lay-in fixtures currently produce hard metric sizes.

Lithonia Lighting, Conyers, GA, one of the largest domestic producers of lay-in fluorescent fixtures, produces hard metric fixture sizes in its SP, SP(air), Paramax, and Optimax product lines. Main number: 404-922-9000. West Coast Contact: Marcus Cone, 818-965-0711.

Columbia Lighting, Spokane, WA, the second largest domestic producer of fluorescent light fixtures, currently sells hard metric fixtures and can produce almost any size metric fixture. Contact 1: Mark Johnson, Product Manager, 509-924-7000. Contact 2: Fred Smith, Product Manager, 509-924-7000.

Some firms that indicate metric production capability and willingness to bid hard metric jobs are listed below:

Midwest Chandelier, Tom Lefkovitz, 913-281-1100.

Lightolier, Mark Wellnitz, 201-864-3000.

Metalux Lighting, Call 912-924-8000.

Day-Brite Lighting, Call 601-842-7212.

Morlite Company, S. Weymouth, MA. Contact: Bob Stacey, 617-331-7732.

Louisville Lamp Company, Louisville, KY. Contact: William Vertres, 502-964-4094.

Solar Kinetics, Dallas, TX (custom manufacturer). Contact: Sandy McCrea, 214-556-2376.

Masonry

Bricks manufactured to 600 mm grid are smaller than the U.S. Imperial size. As production goes down there are more pieces in the same area. This means the overall material price goes up, producing a more costly facade. In Canada, they are manufacturing a Jumbo Brick (Scot) 290 mm long to help reduce cost.

The metric modular brick is 57 x 90 x 190 mm, which corresponds to 2-1/4 x 3-5/8 x 7-1/2". This is almost exactly equal to the American modular brick size, which is 2-1/4 x 3-5/8 x 7-5/8". Since it is only a matter of cutting the brick slightly shorter, many companies can easily make this product.

Glen Gery Corp., Wyomissing, PA, can make metric modular brick. Contact: Ron Hunsicker, Baltimore, 301-837-3170.

Hagerstown Block, Hagerstown, MD, can make metric block. Call 301-733-3510.

Ochs Brick and Tile, Springfield, MN, can make metric modular brick. Contact 1: Rod Schutt, Plant Mgr., 612-944-1450. Contact 2: Bob Larson, Sales Mgr., 612-944-1450.

Plasticrete, North Haven, CT, can make metric block. Contact: Joe Rescigno, 800-243-6934.

U.S. Brick, Streetsville, ON, Canada, has 12 plants in the U.S. that can make metric modular brick. Contact: Ron Spencer (Ontario), 416-821-8800.

Since there are many U.S. brick manufacturers, check with your usual supplier to see if they can make metric modular brick.

Raised Access Flooring

Virtually all manufacturers are currently making 600 x 600 mm grid.

C-TEC, Inc., Grand Rapids, MI, makes a 600 x 600 mm access flooring product line called the Metric Panel. Contact: Don Heeney, 616-243-2211.

Interstitial Systems, Oakbrook, IL, currently manufactures a hard metric 600 x 600 mm raised floor system. Contact: Bill Collier, 708-691-8600.

Appendix B

Tate Architectural Products, Jessup, MD, maker of Tate Raised Access Floors and a world leader in the market, currently produces hard metric 600 x 600 mm raised access flooring systems in its Maryland plant. Systems are available in light, medium, and heavy duty ratings. Heights available from 150 to 900 mm. Contact 1: Lida Poole, 301-799-9123. Contact 2: Victor Sainato, 301-799-9123.

USG Interiors/Donn, Chicago, IL, one of the world's largest commercial interior products companies and a major manufacturer of access flooring, regularly makes hard metric access flooring. Contact 1: William E. Nelson, 312-606-5358. Contact 2: David C. Vanosdall, 312-606-3804.

New Jersey Steel, Sayreville, NJ. Contact: Gary Giovannetti, VP Sales/Marketing, 908-721-6600.

Nucor Steel, Plymouth, UT. Contact: R. Wayne Jones, Sales Mgr., 801-458-3961.

Roofing

While there is no difficulty using customary inch size products in metric construction, roofing products are available in hard metric sizes.

U.S. Intec, Port Arthur, TX, the largest U.S. producer of modified roofing material and the world's largest producer of APP modified bitumen membranes, manufactures one meter by 10 meter rolls of modified membrane in its U.S. plants. Contact 1: Robert Solise, 800-624-6832, ext. 917. Contact 2: Jerry Saunders, 800-624-6832, ext. 915.

Steel Fabrication

Many firms are capable of fabricating steel from metric design drawings. Some of these firms are:

Falcon Steel, Wilmington, DE. Call 302-571-0890.

Havens Steel, Kansas City, MO. Call 816-231-5724.

Interstate Iron Works, White Horse, NJ. Contact: Robert Aberson, 908-534-6644.

Lehigh Structural Steel, Lancaster, SC. Contact: 803-286-5656.

Montague-Betts, Lynchburg, VA. Contact: 804-522-3331.

Steelco Division, Metropolitan Steel, Sinking Spring, PA. Contact: Ron Keating, 215-678-6411.

Suspended Ceiling Systems

There are manufacturers in the U.S. that make 600 mm grid systems. The extruded metal grid system should fall in line rapidly as demand picks up.

Armstrong World Industries, Lancaster, PA, a major SCS manufacturer, currently manufactures and sells hard metric size ceiling products. Except for selected specialty items, the major portion of the Armstrong product line has already been or can be manufactured in metric dimensions. There generally are neither cost premiums nor additional lead times for hard metric products. Contact 1: Dan Kennard, 717-396-2684. Contact 2: Deb Kantner, 717-396-3045.

Capul Architectural Acoustics, Plainfield, IL, a medium-size manufacturer, can produce and bid hard metric projects. Contact: Tom Stanton, Judkins Assoc., Baltimore, MD, 410-234-0010.

Celotex Corp., Tampa, FL, a major manufacturer of SCS, offers an entire product line of hard metric sizes. Contact: George Mitchell, 813-873-4027.

Chicago Metallic Corp., Chicago, IL, a major supplier of SCS grids, produces hard metric grids. Contact: Craig Trotier, 800-323-7164.

National Rolling Mills, Frazier, PA, one of the largest domestic manufacturers of suspension ceiling grids, regularly makes hard metric sizes. Contact: Rich Mattioni, 215-644-6700.

USG Interiors, Chicago, IL, the world's largest commercial interior products company and a major manufacturer of ceiling systems, regularly makes hard metric

Appendix B

size ceiling systems. Contact 1: William E. Nelson, Director, Account Development, 312-606-5358. Contact 2: David C. Vanosdall, Director, Marketing Programs, U.S. Operations, 312-606-3804.

Systems Furniture

Systems furniture products will not need to be converted to hard metric sizes immediately. Many companies export their English-dimensioned products to countries that construct buildings in metric. These products will need their dimensions identified in metric units in product literature. Listed below are some firms that have product literature with metric dimensions.

GF Furniture Systems, Inc., Youngstown, OH, currently exports its English-size panels all over the world. These are successfully utilized in full metric construction. Product literature is available with metric dimensions. Contact: Don Detweiler, 216-533-7799.

Herman Miller, Inc., Zeeland, MI, a major manufacturer of systems furniture, produces both English-size and hard metric size systems furniture. Standard hard metric panels are 600, 800, 1000, 1200, and 1600 mm. Both the metric and the English sizes are sold overseas and utilized in metric construction. All literature is available with metric dimensions. Contact: Mark DeSchon, 616-772-3300

Steelcase, Grand Rapids, MI, a major manufacturer of systems furniture, currently exports its English-size products all over the world. These are successfully used in metric construction. Product literature is available with metric dimensions. Contact: Ken Gilpin, 616-246-4990.

Structural Steel

The AISC has already done the soft conversion to fit into existing shapes and strengths. Due to the variety of structural shapes, this soft conversion, in all probability, will remain.

Several major wide flange steel beam manufacturers currently produce hard metric steel beams to the Japanese and German standards:

Chaparelle Steel, Midlothian, TX. Contact: Jim Wroble, 800-527-7979.

Nucor Steel, Blytheville, AR produces only some metric shapes. Contact: Bob Johns, 800-289-6977.

All LRFD property and specification design data will be available in metric from AISC in January 1992, for standard American A6/A6M steel shapes.

AISC recommends using standard A6/A6M shapes for metric construction in the United States.

Temperature Controls

All of the major manufacturers of HVAC controls currently offer products that will operate in Celsius or Fahrenheit. Some of those firms are: Johnson Controls, Barber Colman, Robertshaw, Andover, and Honeywell. Contact your local representative for ordering information.

Tools

Measurement-sensitive tools are available from domestic manufacturers.

Stanley manufactures tape measures and rulers for the international market. They will probably change weights and sizes to metric in soft conversion first.

Stanley Tools, New Britain, CT, domestically manufactures metric and metric/English tape measures: Model 32-158, Metric/English, 5 m / 16 ft.; Model 32-156, Metric, 5 m; Model 33-428, Metric/English, 7.5 m / 25 ft.; Model 33-443, Metric, 10 m. The 33-443 is a new introduction and may not be available. Check with Stanley for availability.

Appendix B

Contact 1: Carl Lickwar, 203-225-5111. Contact 2: Alan Martin, 203-225-5111. These products can be ordered directly from Stanley or through your local hardware store.

Metric scales are available from:

Staedtler-Mars: Model 987-18-1 Alvin, Model 117 PM. Contact your local graphic arts supply store to order scales. Note: These metric scales are made overseas, as are most inch-size architect scales available today.

Windows

Commercial window systems are available in hard metric sizes.

Alenco Commercial Group, Bryan, TX, one of the largest U.S. manufacturers of aluminum windows, currently makes metric windows primarily for export, and can make any size for domestic use. Contact: Harold Chilton, 409-823-6557.

Andersen Windows, Commercial Group, Bayport, MN, currently fabricates windows in its one domestic plant and exports to several countries. Product literature is available with metric dimensions. Contact: Craig Johnson, 612-439-5150.

Caradco, IL, can make any size hard metric window. Contact: Roy Szyhowski, 217-893-4444.

Desco Company, Desmet, SD, can produce hard metric sizes. Contact: Cindy Albrecht, 605-854-9126.

Howard Industries, Miami, FL, one of the nation's leading producers of energy saving windows, doors, and windowwall systems, has worked on foreign projects before and can bid many metric projects. Contact 1: Bob Boigt, 305-888-1521. Contact 2: Joe Sixto, 305-888-1521.

Marmet Corp., Wausau, WI, can make any size metric window. Contact: Brent Schepp, 715-845-5242.

Marvin Windows, Warroad, MN, has previously manufactured and can produce windows in metric sizes. Contact: Dan McKinnon, 218-386-1430.

Optimum Windows, Bronx, NY, can produce hard metric sizes. Contact: Candido Perex, 212-991-0700.

Peerless Commercial Window Division, Kansas City, MO, can make any size hard metric window. Contact: Tony Grossi, 913-432-2232.

Pella Windows, Pella, IA, can make any size metric window. Contact: Julio Chiarella, 515-628-1000.

Appendix C
Metric References

Metric Construction Guides

The following are reference sources for further information on general and specific metric topics. Price and availability were accurate at the time this book was printed, but should be verified before ordering.

Much of the information in this section is from "Metric In Construction," a bimonthly newsletter published by the Construction Metrication Council. For more information about the Council, write to the Construction Metrication Council, National Institute of Building Sciences, 1201 L St., N.W., Suite 400, Washington, D.C. 20005, or call (202) 289-7800.

National Institute of Building Sciences (Publications Department, 1201 L St., N.W., Suite 400, Washington, DC 20005; phone 202-289-7800):

- *Metric Guide for Federal Construction*. Written by NIBS specifically for the construction industry and reviewed by metric experts throughout the country. Includes background on federal metric laws; facts on metric in construction; an introduction to metric units; a primer on metric usage for architects, engineers, and the trades; requirements for metric drawings and specifications; guidance on metric management and training; and a list of current metric construction references. 34 pp. $15.00 (including shipping and handling).
- *GSA Metric Design Guide*. Interim design guide developed by the General Services Administration (GSA) for use by federal project managers and their A/Es. Contains practical architectural, civil, structural, mechanical, and electrical design information; a list of available "hard" metric products; sample drawings; and related reference information. 77 pp. $8.00; $5.00 if ordered with the above *Metric Guide for Federal Construction*.

General Metric Information

American National Metric Council (Washington, DC; phone 410-727-0882):
- *ANMC Metric Editorial Guide*. $5.00; bulk discounts available.
- *SI Metric Training Guide*. $5.00
- *Metrication and the Consumer*. $5.00
- *Metrication for the Manager*. $15.00

U.S. Metric Association (10245 Andasol Ave., Northridge, CA 91325; phone 818-363-5606):

- *Style Guide to the Use of the Metric System*. $3.00; bulk discounts available.

Appendix C

- *SI Metric Style Guide for Written and Computer Usage.* $2.00; bulk discounts available.
- *Freeman Training/Education Metric Materials List.* $38.00.
- *Metric Vendor List.* $28.00.

Blackhawk Metric Supply Inc. (Box 543, South Beloit, IL 61080; phone 815-389-2850):

- *Metric poster.* Attractive four-color chart that shows common metric units and their logical relationships. $8 each; bulk discounts available.

Design

American Institute of Architects (AIA Bookstore, 1735 New York Ave., N.W., Washington, DC 20006; phone 202-626-7475. All but the *AIA Pocket Metric Guide* are published by John Wiley & Sons, Professional Reference and Trade Group, 605 Third Ave., New York, NY 10158; phone 1-800-225-5945, ext. 2497):

- *AIA Pocket Metric Guide.*
- *Architectural Graphic Standards.* A metric edition is not due for several years, but current editions include a comprehensive section on metric conversion.
- *The Architect's Studio Companion: Technical Guidelines for Preliminary Design.* By Edward Allen and Joseph Iano. Includes dual units. $52.95.
- *Architectural Detailing: Function, Constructability, and Aesthetics.* By Edward Allen. Includes dual units. Available with *Architect's Studio Companion* as a set for $87.50.
- *Fundamentals of Building Construction: Materials and Methods.* By Edward Allen. Includes dual units. $59.95.
- *Neufert Architect's Data.* By Ernst Neufert. Second International (metric) Edition (Germany). $52.50.
- *Wiley Engineer's Desk Reference.* By S.I. Heisler. Includes dual units. $54.95.

Cost Estimating

R. S. Means Company. (Box 800, Kingston, MA 02364; phone 617-585-7880):

- *Means Building Construction Cost Data, Metric Version* (updated annually). $94.95 for the 1993 edition. Contact Means for information about metric cost data available electronically through MeansData™ for Spreadsheets products, and from MeansData™ authorized resellers.

Specifications

American Institute of Architects (1735 New York Ave., N.W., Washington, DC 20006; phone 1-800-424-5080):

- *AIA MASTERSPEC* will contain dual units as of 1993.

Construction Specifications Institute (601 Madison St., Alexandria, VA 22314-1791; phone 703-684-0300):

- *CSI SPECTEXT* contains dual units. All other CSI publications contain dual units or are in the process of being converted.

Building Codes

Building Officials and Code Administrators International (4051 W. Flossmoor Rd., Country Club Hills, IL 60477-5795; phone 312-799-2300):

- *BOCA National Building, Fire Prevention, Mechanical, and Plumbing Codes.* All editions are published with dual units.

International Conference of Building Officials (5360 South Workman Mill Rd., Whittier, CA 90601; phone 310-699-0541):

Appendix C

- *Uniform Building, Fire, Mechanical, and Plumbing Codes.* The 1994 editions will be published with dual units.

National Fire Protection Association (1 Batterymarch Park, Box 9101, Quincy, MA 02269-9101; phone 1-800-344-3555):

- NFPA 101, *Life Safety Code.* All NFPA documents are published with dual units. $33.75.

Southern Building Code Congress International, Inc. (900 Montclair Rd., Birmingham, AL 35213-1206; phone 205-591-1853):

- *Standard Building Code.* The 1991 edition is published with dual units. The *Standard Fire, Plumbing, and Mechanical Codes* will be published with dual units in 1994.

Standards

American National Standards Institute, Inc. (11 W. 42nd St., New York, NY 10036; phone 212-642-4900):

- ANSI/IEEE 268, *American National Standard Metric Practice.* $52.50.
- ANSI/AWS A1.1, *Metric Practice Guide for the Welding Industry.* $20.00.
- ANSI/IEEE 945, *Preferred Metric Units for Use in Electrical and Electronics Science and Technology.* $45.00.
- ISO 1000, *SI Units and Recommendations for the Use of Their Multiples and Certain Other Units.* $48.00.
- Many other ANSI standards are available in metric; check with ANSI.

American Society for Testing and Materials (1916 Race St., Philadelphia, PA 19103; phone 215-299-5585):

- ASTM E380, *Standard Practice for Use of the International System of Units (SI).* $23.00.
- ASTM E621, *Standard Practice for the Use of Metric (SI) Units in Building Design and Construction.* $23.00
- ASTM E713, *Guide for Selection of Scales for Metric Building Drawings.* $15.00
- ASTM E577, *Guide for Dimensional Coordination of Rectilinear Building Parts and Systems.* $15.00.
- ASTM E835, *Guide for Dimensional Coordination of Structural Clay Units, Concrete Masonry Units, and Clay Flue Linings.* $15.00.
- All other ASTM standards are published in metric or with dual units.

Underwriters Laboratories, Inc. (355 Pfingston Rd., Northbrook, IL 60062; phone 708-272-8800):

- Virtually all UL standards contain dual units.

Civil

American Congress on Surveying and Mapping (5410 Grosvenor Lane, Suite 100, Bethesda, MD 20814; phone 301-493-0200):

- *Metric Practice Guide for Surveying and Mapping.* $10.00.

American Association of State and Highway Transportation Officials (444 N. Capital St., N.W., Suite 225, Washington, DC 20001; phone 202-624-5800):

- *Guide to Metric Conversion. Standard Specifications for Transportation Materials.* Two-volume set. Includes dual units. $115.00.

Wood

American Forest and Paper Association (formerly National Forest Products Association; 1250 Connecticut Ave., N.W., Washington, DC 20036; phone 202-463-2700):

Appendix C

- *Lumber and Wood Products Metric Planning Package.* Currently being revised.

National Particleboard Association (18928 Premiere Ct., Gaithersburg, MD 20879; phone 301-670-0604):

- Metric units currently are being added to the APA/ANSI standards for particleboard and medium-density fiberboard.

Hardwood Plywood Manufacturers Association (Box 2789, Reston, VA 22090-2789; phone 703-435-2900):

- *Interim Voluntary Standard for Hardwood and Decorative Plywood.* Includes dual units. $10.00.

Steel

American Institute of Steel Construction (Metric Publications, 1 E. Wacker Dr., Suite 3100, Chicago, IL 60601-2001; phone 312-670-5414):

- *Metric Properties of Structural Shapes with Dimensions According to ASTM A6M.* Metric version of Part I of the *Manual of Steel Construction.* $10.00.
- *Metric Conversion: Load and Resistance Factor Design Specificaton for Structural Steel Buildings.* $10.00.
- *Manual of Steel Construction, Metric Edition.* To be published in 1994.

American Welding Society (550 N.W. Le Jeune Rd., Box 35104, Miami, FL 33135; phone 305-443-9353):

- All AWS standards include dual units.

Industrial Fasteners Institute (1105 East Ohio Building, 1717 E. 9th St., Cleveland, OH 44114; phone 216-241-1482):

- *Metric Fastener Standards.* $60.00

Concrete

American Concrete Institute (Box 19150, Detroit, MI 48219; phone 313-532-2600):

- ACI 318M-89/318RM-89, *Building Code Requirements for Reinforced Concrete and Commentary.* Metric edition of ACI 318-89/318R-89. $70.00.
- ACI 318.1M-89/318.1RM-89, *Building Code Requirements for Metric Structural Plain Concrete and Commentary.* Metric edition of ACI 318.1-89/318.1R-89. $11.50.

Mechanical and Electrical

American Society of Heating, Refrigerating, and Air Conditioning Engineers (1791 Tullie Circle, N.E., Atlanta, GA 30329; phone 404-636-8400):

- *SI for HVAC & R.* Free on request.
- *Psychrometric Charts SI.* Charts 1 through 7. $10.00.
- *1991 Handbook – HVAC Applications.* SI edition. $114.00.
- *1989 Handbook – Fundamentals.* SI edition. $114.00.
- *1990 Refrigeration Handbook.* SI edition. $114.00.
- *1992 Handbook – HVAC Systems and Equipment.* SI edition. $114.00.
- All ASHRAE standards are published in metric or with dual units. ASHRAE plans to discontinue the use of inch-pound units by the year 2000.

American Society of Mechanical Engineers (22 Law Dr., Box 2300, Fairfield, NJ 07007; phone 1-800-843-2763 ext. 426):

- SI-1, *Orientation and Guide for Use of SI (Metric) Units.* $12.00.
- SI-2, *SI Units in Strength of Materials.* $12.00.
- SI-3, *SI Units in Dynamics.* $12.00.

Appendix C

- SI-4, *SI Units in Thermodynamics*. $12.00.
- SI-5, *SI Units in Fluid Mechanics*. $12.00.
- SI-6, *SI Units in Kinematics*. $12.00.
- SI-7, *SI Units in Heat Transfer*. $12.00.
- SI-8, *SI Units in Vibration*. $12.00.
- SI-9, *Guide for Metrication of Codes and Standards Using SI (Metric) Units*. $13.00.
- SI-10, *Steam Charts, SI (Metric) and U.S. Customary Units*. Edited by J. H. Potter. $28.00
- All other ASME standards, except the *Boiler and Pressure Vessel Code*, are published either in separate SI editions or with dual units.

National Environmental Balancing Bureau (1385 Piccard Dr., Rockville, MD 20850; phone 301-977-3698):

- *Fundamentals*, *Air System*, and *Hydronic Systems* guides. Available in metric editions.

National Fire Protection Association (1 Batterymarch Park, Box 9101, Quincy, MA 02269-9101; phone 1-800-344-3555):

- NFPA 13, *Installation of Sprinkler Systems*. Includes dual units. $24.50.
- ANSI/NFPA 70, *National Electrical Code*. Includes dual units. $32.50.
- All other NFPA standards are published with dual units.

Sheet Metal and Air Conditioning Contractors National Association (4201 Lafayette Center Dr., Chantilly, VA 22021; phone 703-803-2980):

- All SMACNA publications are being converted to dual units.

Water Environment Federation (601 Wythe St., Alexandria, VA 22314; phone 703-684-2400):

- Manual of Practice No. 6, *Units of Expression for Wastewater Treatment Management*. $15.00.

Product Manufacturing

Association for Manufacturing Technology, The (7901 Westpark Dr., McLean, VA 22102-4269; phone 703-893-2900):

- *Guidelines for Metric Conversion in Machine Tool and Related Industries.* $15.00.

Metric Calculator

Sharp Instrument Company (Van Schaack Premium Group, 4747 W. Peterson, Chicago, IL 60646; phone 312-736-5600):

- *Sharp Model EL-344G Metric Calculator*. Converts linear dimensions, areas, volumes, liquids, pressures, and masses with two keystrokes. Very handy for simple conversions. Under $20.00.

Metric Conversion Software

Orion Development Corporation (Box 2323, Merrifield, VA 22116-2323; phone 1-800-992-8170):

- *Metric-X*. Metric conversion software for use with any IBM-compatible computer. Single user copies $24.95; site/network copies available; bulk discounts available.

Vidtrack Technologies Co. (540 S. Main St., Suite 941, Akron, OH 44311-1010; phone 216-762-5141):

- *ConvertFile Conversion Utility*. Metric conversion software for use with IBM-compatible computers. $29.95.

MCB Enterprises (Box 6563, Huntington Beach, CA 92615-6563; phone 714-647-5534):

Appendix C

- *Metric Calc!* Metric conversion software in both DOS and Windows versions. $49.95.

CAD and Engineering Software

The two largest CAD vendors, **Autodesk** and **Intergraph**, allow the user to work in either inch-pound or metric units. Many structural and mechanical design programs have metric capability, too. Before you purchase any new computer software, **make sure it has provisions for metric**.

Metric Video Tapes

Workplace Training (520 N. Arm Dr., Orono, MN 55364; phone 612-472-2564):

- *SI Metric for the Workplace.* Six-tape video/workbook courseware. Provides in-depth metric training for business and industry professionals. Includes an introduction to metric, units of measure, reading/writing rules, limits/fits/tolerances, and metric conversion. $2195.

MMEI Corporation (2247 Lexington Pl., Livermore, CA 94550; phone 510-449-8992):

- *All About Metric.* Three-tape video training package that covers the background of the metric system, government/industry transition, everyday metric units, and rules for metric usage. Accompanying reference manual includes conversion tables and other information. Instructor's manual contains lesson tips, test questions, illustrations suitable for use as masters for overhead projector transparencies. $500. Also available through the U.S. Metric Association.

Appendix D
Professional Associations

ACRI   Air-Conditioning and Refrigeration Institute
1501 Wilson Blvd., 6th Floor
Arlington, VA 22209

AMCA   Air Movement and Control Association
30 West University Drive
Arlington Heights, IL 60004

AA   Aluminum Association
900 19th Street, NW, Suite 300
Washington, DC 20006

AACE   American Association of Cost Engineers
209 Prairie Ave., Suite 100
Morgantown, WV 26505

AAN   American Association of Nurserymen, Inc.
1250 I St., NW, Suite 500
Washington, DC 20005

ACI   American Concrete Institute
Box 19150
Redford Station
Detroit, MI 48219

ACEC   American Consulting Engineers Council
1015 15th Street, NW
Washington, DC 20005

Appendix D

AIA	American Institute of Architects 1735 New York Avenue, NW Washington, DC 20006
AIC	American Institute of Constructors 9887 N. Gandy, Suite 104 St. Petersburg, FL 33702
AISC	American Institute of Steel Construction 1 E. Wacker Dr., Suite 3100 Chicago, IL 60601
AITC	American Institute of Timber Construction 11818 SE Mill Plaine Blvd., Suite 415 Vancouver, WA 98684
AISI	American Iron and Steel Institute 1133 15th Street, NW Washington, DC 20005
ANSI	American National Standards Institute 11 West 42nd St. New York, NY 10036
APA	American Plywood Association Box 11700 Tacoma, WA 98411
ASCE	American Society of Civil Engineers 345 East 47th Street New York, NY 10017
ASHRAE	American Society of Heating, Refrigerating and Air Conditioning Engineers 1791 Tullie Circle, NE Atlanta, GA 30329
ASLA	American Society of Landscape Architects 4401 Connecticut Ave., NW Washington, DC 20008
ASME	American Society of Mechanical Engineers 345 East 47th Street New York, NY 10017
ASPE	American Society of Professional Estimators 11141 Georgia Ave, Suite 412 Wheaton, MD 20902

Appendix D

ASTM	American Society for Testing and Materials 1916 Race Street Philadelphia, PA 19103
AWWA	American Water Works Association 6666 West Quincy Avenue Denver, CO 80235
AWS	American Welding Society 550 Lejeune Rd, NW Miami, FL 33126
AWC	American Wood Council 1250 Connecticut Avenue, NW, Suite 300 Washington, DC 20036
AWPA	American Wood Preservers' Association P.O. Box 286 Woodstock, MD 21163
AWI	Architectural Woodwork Institute P.O. Box 1550 Centerville, VA 22020
AI	Asphalt Institute Research Park Dr. P.O. Box 14052 Lexington, KY 40512
ABC	Associated Builders and Contractors, Inc. 729 15th Street, NW Washington, DC 20005
AED	Associated Equipment Distributors 615 W. 22nd Street Oak Brook, IL 60521
AGC	Associated General Contractors of America 1957 E Street, NW Washington, DC 20006
ALCA	Associated Landscape Contractors of America 405 N. Washington Steet Falls Church, VA 22046

Appendix D

BIA	Brick Institute of America 11490 Commerce Park Dr., Suite 300 Reston, VA 22091
BHMA	Builders Hardware Manufacturers Assoc. 355 Lexington Ave., 17th Floor New York, NY 10017
CFMA	Construction Financial Management Association 40 Brunswick Ave., Suite 202 Edison, NJ 08818
CIMA	Construction Industry Manufacturers Assoc. 111 E. Wisconsin Ave., Suite 940 Milwaukee, WI 53202
CMAA	Construction Management Association of America 1893 Preston White, Suite 130 Reston, VA 22091
CSI	Construction Specifications Institute 601 Madison Street Alexandria, VA 22314
CRSI	Concrete Reinforcing Steel Institute 933 Plum Grove Road Schaumberg, IL 60173
CDA	Copper Development Association, Inc. P.O. Box 1840 Greenwich, CT 06836
DHI	Door & Hardware Institute 14170 Newbrook Drive Chantilly, VA 22021
FMERO	Factory Mutual Engineering & Research Group 1151 Boston-Providence Turnpike Norwood, MA 02062
FGMA	Flat Glass Marketing Association White Lakes Professional Building 3310 SW Harrison Street Topeka, KS 66611

Appendix D

GSA	General Service Administration F Street and 18th, NW Washington, DC 20405
GA	Gypsum Association 810 1st Street, NE, Suite 510 Washington, DC 20002
IEEE	Institute of Electrical and Electronic Engineers 345 East 47th Street New York, NY 10017
IMI	International Masonry Insitute 823 15th Street, NW Washington, DC 20005
MCAA	Mechanical Contractors Association of America, Inc. 1385 Piccard Drive Rockville, MD 20832
MLSFA	Metal Lath/Steel Framing Association 600 South Federal, Suite 400 Chicago, IL 60605
NAAMM	National Association of Architectural Metal Manufacturers 600 South Federal, Suite 400 Chicago, IL 60605
NAHB	National Association of Home Builders 15th & M Street, NW Washington, DC 20005
NAPHCC	National Association of Plumbing-Heating- Cooling Contractors P.O. Box 6808 Falls Church, VA 22046
NECA	National Electrical Contractors Association 7315 Wisconsin Avenue Bethesda, MD 20814

Appendix D

NEMA	National Electrical Manufacturers Association 2101 L Street, NW, Washington, DC 20037
NIBS	National Institute of Building Sciences 1201 L Street NW, Suite 400 Washington, DC 20005
NFPA	National Fire Protection Association One Batterymarch Park Quincy, MA 02269
NFPA	National Forest Products Association 1250 Connecticut Avenue, NW Washington, DC 20036
NPCA	National Precast Concrete Association 825 E. 64th Street Indianapolis, IN 46220
NRCA	National Roofing Contractors Association 10255 West Higgins Road, Suite 600 Rosemont, IL 60018
NSPE	National Society of Professional Engineers 1420 King Street Alexandria, VA 22314
NSWMA	National Solid Wastes Management Association 1730 Rhode Island Avenue, NW Washington, DC 20036
NUCA	National Utility Contractors Association 1235 Jefferson Davis Hwy., Suite 606 Arlington, VA 22202
PCA	Portland Cement Association 5420 Old Orchard Road Skokie, IL 60077
PCEA	Professional Construction Estimators Association P.O. Box 1107 Cornelius, NC 28031

Appendix D

PS Product Standard
U.S. Department of Commerce
14th & Constitutional Avenue
Washington, DC 20230

SIGMA Sealed Insulating Glass Manufacturers Association
401 N. Michigan Ave.
Chicago, IL 60601

SMACNA Sheet Metal and Air Conditioning Contractors
    National Association, Inc.
4201 Lafayette Center Drive
Chantilly, VA 22021

SDI Steel Door Institute
30200 Detroit Road
Cleveland, OH 44145

SSPC Steel Structures Painting Council
4400 Fifth Avenue
Pittsburgh, PA 15213

TCA Tile Council of America, Inc.
P.O. Box 326
Princeton, NJ 08542-0326

UL Underwriters Laboratories, Inc.
333 Pfingsten Road
Northbrook, IL 60062

USFS U.S. Forest Service
Forest Products Laboratory
P.O. Box 5130
Madison, WI 53705

WCLIB West Coast Lumber Inspection Bureau
6980 SW Varns Street
Box 23145
Portland, OR 97223

WWPA Western Wood Products Association
Yeon Bldg.
522 SW 5th Ave.
Portland, OR 97204

Glossary of Terms

Basic module
The fundamental unit of size in the systems of coordination in metric building, or 100 mm. Designated by the symbol M.

Basic units of the metric system
Meter, kilogram, liter, degree Celsius.

Degree Celsius
Unit of temperature (°C). 0° Celsius is equivalent to 32° F (the freezing point of water), and 100° Celsius is equivalent to 212° F (the boiling point of water). A Celsius degree is 1.8 times larger than a Fahrenheit degree.

Dimensional coordination
Special dimensional preferences for buildings and building products. In metric dimensional coordination, a common set of preferred dimensions is used to establish the geometry of buildings as well as the sizes of constituent components or assemblies. (Also called *modular coordination*).

Dual dimensions
The expression of dimensions in both customary and metric units of measure.

Dual labeling
The use of both customary and metric designations on labels.

English Imperial measurement system
The foot, pound, gallon, degree Fahrenheit system presently used in the U.S.

Estimate
A reliable cost/time evaluation of an item or project both in part and in total, for both present and life cycle.

Hard conversion
Changing the actual size of a product so that its measurements are in rounded metric sizes.

Inframodular size
A selected dimension smaller than the basic module (100 mm).

Intermodular size
A selected dimension larger than the basic module (100 mm), but not a whole multiple of the module.

Kilogram
Base unit of mass (weight). Symbol: kg. The kilogram is equivalent to 1000 grams and is approximately 2.2 pounds.

Liter
Unit of volume or capacity used mainly to measure quantities of liquid or gaseous materials. It is equal to a cubic decimeter. Symbol: L. The liter is 6% larger than a quart. A smaller unit is the milliliter (mL), which is 1/1000 of a liter.

Meter
The base unit of length. Symbol: m. The meter is equivalent to 39.37 inches. Smaller units are the centimeter (cm), which is 1/100 of a meter (a little less than 1/2 inch), and the millimeter (mm), which is 1/1000 of a meter (a little more than

Glossary

1/32 inch). The kilometer (km) is a larger unit and is equivalent to 0.62 mile.

Metrication
The process of converting to the metric system.

Preferred multimodular dimension
A selected multiple of the basic module (100 mm).

Rationalization
The process of adopting sizes based on logical increments that minimize the variety in inventory stocks.

Rounded metric sizes
Sizes expressible in simple numbers such as 500 g, 1 kg, 2 kg, 500 mL, 1 L, 2 L, etc.

SI
Abbreviation for the modern metric system, the International System of Units (from the French, Systeme International d'Unites). It evolved from the original French metric system, and is currently being used by most nations of the world.

Soft conversion
The translation of customary unit measurements to their equivalent values in metric units.

Triangulation
Common method used to convert irregular areas into square meters by dividing the areas into a series of triangles and rectangles or squares.

Index

A
ACI (American Concrete Institute), 169
APA (American Plywood Association), 170
Area conversions, 146
Architect's First Source, 107
Architectural equipment, 109
ASHRAE (American Society of Heating Refrigerating & Air Conditioning Engineers), 170
ASTM (American Society for Testing and Materials), 171

B
Base units, Symbols for, 5
BIA (Brick Institute of America), 69-71, 172
Brick
  common, 70
  dimensions, shapes, and joints, 74
  sizes, 70
Building construction, dimensions preferred for, 8-16

C
CAD and engineering software, 168
Changes required for metric, 115
Civil and structural engineering, conversion factors, 10
Concrete
  block, 75-77
    masonry, 75
  costs, 55
Consolidated estimate form, 27-28
Constructability, 17
Construction
  metric units used in, 13
  specialties, 107-108
Construction Metrication Council, 133
Construction Specifications Institute, 39
Conversion and rounding, 6
Conversion factors
  civil and structural engineering, 10
  electrical engineering, 12
  mechanical engineering, 11

Conveying Systems, 115
Coordination, modular, 8
Cost Analysis Sheet, 24
Costs, types of, 23
  equipment, 25-26
  labor, 25
  material, 23, 25
    factors to analyze, 23
CSI (Construction Specifications Institute), 172
CSI MasterFormat, 39

D
Design
  metric considerations in, 133
Dimension coordination, 8
Division 1, General Conditions, 41
  changes required for metric, 41
  estimating procedures, 41
    cleanup, 43
    contract provisions, 42-43
    equipment and general charges, 43
    labor provisions, 43
    office costs, 43
    personnel, 41-42
    services, 42
  project overhead summary, 44-46
Division 2, Site Work, 47
  changes required for metric, 47-48
    AASHTO (American Association of State Highway and Transportation Officials), 47
    FHWA (Federal Highway Administration), 48
    rebar, 47
  clear and grub, 49
  equipment size, 50
  estimate sheets, 51-53
  estimating procedures, 48
  grading, 50
  preparing the estimate, 48
  sample takeoff, 50-53
  site clearing, 49

179

site exploration, 48
swell and shrinkage, 50
Division 3, Concrete, 55
  beams, elevated, 61-62
  building slab on grade, 64
  cast-in-place slabs, 64
  cementitous decks, 67
    quantity takeoff, 67
  changes required for metric, 55
    ACI (American Concrete Institute), 55
    concrete, 55
    rebar, 55
  columns, 60
  continuous footing, 57
  curbs and gutters, 65
  elevated slab, 64
  estimating procedures, 56
    cast-in-place formed concrete, 56
  forms and shoring, 65
  foundation detail, 59
  other finished surfaces, 64
  piers, 58
  pile caps, 58
  placing, 66
  precast, 67
  pricing
    cast-in-place, 62, 66-67
      placing, 62
      reinforcing, 62
    cementitous decks, 67
    precast, 67
    other finished surfaces, 66-67
  reinforcing bars, 63
    imperial and metric sizes and weights, 63
  sample project takeoff, 68
  stairs, 65
  underpinning, 61
  walls, 58-59
Division 4, Masonry, 69
  changes required for metric, 69
    BIA (Brick Institute of America), 70-71
    brick, 70-71
    concrete block, 70-71
    NCMA (National Concrete Masonry Association), 71
  estimating procedures, 71
    bonding, 72-73
    concrete block masonry, 75
    lintels, 75
    measurements, 71-72
    mortar, 72
  sample project takeoff, 78
Division 5, Metals, 79
  basic components, 79
    decks, 79
    expansion control, 79
    fasteners, 79
    joists, 79
    metals, 79
    miscellaneous, 79
    ornamental, 79
    structural, 79
  changes required for metric, 79

    AISC (American Institute of Steel Construction), 79
    structural steel, 79
  estimating procedures, 80
    erection, 81
    miscellaneous metals, 80
    ornamental iron, 80
    steel deck, 80
    steel joists, 80
    structural steel, 80
  sample project takeoff, 82
Division 6, Wood and Plastics, 83
  changes required for metric, 83-84
    wood product associations on metric, 83
  estimating procedures, 84
    finish carpentry, 86-87
    rough carpentry, 84, 86
  sample project takeoff, 88
Division 7, Thermal and Moisture Protection, 89-93
  changes required for metric, 89
    flashing, 90
    insulation, 89
    NICA (National Insulation Contractors Association), 89
    roofing, 89
    sliding, 90
  estimating procedures, 90
    dampproofing, 90
    insulation, 91
    roof accessories, 92
    roofing and siding, 91
    sheet metal work, 92
    shingles, 91
    single-ply roofs, 91
    waterproofing, 90
  sample project takeoff, 93
Division 8, Doors and Windows, 95-100
  changes required for metric, 95
    associations on metric, 95
    doors, 95
    windows, 95
  estimating procedures, 95
    doors, 97
    entrances and storefronts, 97
    finish hardward and specialties, 98
    glass and glazing, 99
    metal doors and frames, 95
    metal windows, 98
    special doors, 97
    window/curtain walls, 99
    windows, pricing, 98
    wood and plastic doors, 97
    wood windows, 98
  sample project takeoff, 100
Division 9, Finishes, 101
  Changes required for metric, 101
    ceiling system, 101
    drywall, 102
    lathing and plastering, 101
    paint and wallcoverings, 102
    raised floor systems, 101
    tile, terrazzo and flooring, 102

estimating procedures, 102
   acoustical treatment, 103
   drywall takeoff, 102-103
   flooring, 103
   lathing and plastering, 102
   painting, 104
   tile and terrazzo, 103
   wall coverings, 104
sample project takeoff, 105-106
Division 10, Specialties, 107-108
  changes required for metric, 107
  estimating procedures, 107-108
  sample project takeoff, 108
Division 11, Architectural Equipment, 109
  changes required for metric, 109
  estimating procedures, 109-110
Division 12, Furnishings
  changes required for metric, 111
  estimating procedures, 111
Division 13, Special Construction, 113
  changes required for metric, 113
  estimating procedures, 113
Division 14, Conveying Systems, 115
  changes required for metric, 115
Division 15, Mechanical, 117
  changes required for metric, 117
    air diffusers and grilles, 118
    air distribution, 118
    ASHRAE (American Society of Heating, Refrigerating and Air Conditioning Engineers) on metric, 117
    ductwork, 118
    HVAC controls, 119
    mechanical equipment, 119
    pipe and tubing, 119
    sheet metal ductwork, 119
    temperature, 118
  estimating procedures, 119
    general review, 119
    material and labor pricing, 121
    material takeoff, 120
  sample project takeoff, 122-123
Division 16, Electrical, 125-130
  changes required for metric, 125
    lighting fixtures, 125
    NEMA (National Electrical Manufacturers Association) on metric, 126
  estimating procedures, 126-129
    estimate summary, 129
    pricing the estimate, 129
    quantity takeoff, 127
    transfer of material takeoff to price sheets, 127-128
  sample project takeoff, 130
Door and frame schedule, typical, 96
Doors, 95-100
Drawings, 133
  a word of caution, 134
  metric requirements for, 133
  sample metric, 136-142
  scales, metric vs. customary, 134
  slopes, expression of in metric, 134

E
Electrical, 125-130
  NEMA (National Electrical Manufacturers Association)- approved conduit designations, 126
Electrical engineering, conversion factors, 12
Equipment costs, 25-26
  operating, 26
  ownership, 26
Estimate(s)
  considerations, 17
  consolidated, 27
  definition of, 17
  sample
    project plans for, 40
  sample project, 39
Estimating
  common methods of, 18
    order of magnitude, 18
    square meter and cubic meter, 18
    systems, 18
    unit price, 18

F
Finishes, 101
Forms
  consolidated estimate form, 28
  cost analysis sheet, 24
  project overhead summary, 45-46
Furnishings, 111

G
General Conditions, 41-46
  estimating procedures, 41
    cleanup, 43
    contract provisions, 42
    equipment and general charges, 43
    labor provisions, 43
    office costs, 43
    personnel, 41
    services, 42
Glossary of Terms, 177

I
Index, 179
Interior finishes, 101

L
Labor
  costs, 25
  rates, 31-33
Laminated construction, 84-86
Linear conversions, 146
Lintels, 75
Lumber, standard dimensions, 85

M
Masonry, 69
Material costs, 23
Means Building Construction Cost Data, Metric Version, 26, 39, 67, 69, 107, 109, 115
  bare costs, 30

City Cost Index, 31, 35
crews, 30
daily output, 30
descriptions, 29
labor rates, 31-33
location adjustment, 35-36
man-hours, 30
numbering system, 29
square meter and cubic meter costs, 33
time adjustment, 36
total including overhead & profit, 31
unit column, 30
using, 29-38
Mechanical, 117
Mechanical engineering, conversion factors, 11
Metric calculator, 167
Metric Conversion Tables, 146-154
Metric conversion software, 167-168
Metric Estimating by MasterFormat Division, 39-130
Metric, factors for converting, 148-154
Metric measurements, visualizing, 6
Metric product sources, 155
    air diffusers & grilles, 155
    chillers, 156
    curtainwall systems, 156
    custom products, 156
    doors & frames, 156
    drywall, 157
    elevators, 157
    lighting fixtures, 158
    masonry, 158
    raised access flooring, 158
    roofing, 159
    steel fabrication, 159
    suspended ceiling systems, 159-160
    systems furniture, 160
    structural steel, 160
    temperature controls, 160
    tools, 160-161
    windows, 161
Metric references, 163
    building codes, 164-165
    civil, 165
    concrete, 166
    construction guides, 163
    cost estimating, 164
    design, 164
    general information, 163-164
    mechanical and electrical, 166-167
    product manufacturing, 167
    specifications, 164
    standards, 165
    steel, 166
    wood, 165-166
Metric symbols and names, writing, 6
Metric units
    based with prefixes, 146
    used in construction, 13
Metric video tapes, 168
Modular coordination, 8
Module, metric, 8

N
NEMA (National Electrical Manufacturers Association), 174
NIBS (National Institute of Building Services), 174

O
Overhead and profit, 17

P
Professional associations, 169
Preferred dimensions
    inframodular sizes, 9-15
    intermodular sizes, 9-15
    matrix of criteria for, 16
    multimodular dimensions, 9-15
Pricing the estimate, 23-28
    equipment costs, 25-26
    forms, 23
    labor costs, 25
    pricing the estimate, 25
    subcontractors, 26-27
Project Overhead Summary, 45-46

Q
Quantity takeoff, extending the, 22-23
    sample, 22-23
Quantity takeoff procedures, 19-22
    carpentry, 20
    concrete, 19
    doors, windows, storefronts, and finish hardware, 20
    finishes, 20
    guidelines for, 18
    masonry, 19
    mechanical and electrical, 21
    moisture protection, 20
    shortcut methods, 21-22
    site work, 21
    specialities, architectural equipment, furnishings, special construction, and conveying systems, 21
    structural steel, miscellaneous iron, ornamental metals, 19-22
    summary, 21

R
Rules for using metric
    for area, 7
    for linear measurement (length), 7
    for volume and fluid capacity, 7

S
Sample metric drawings, 136-142
Sample project
    description of, 39
    Division 2 takeoff, 51-53
    Division 3 takeoff, 68
    Division 4 takeoff, 78
    Division 5, takeoff, 82
    Division 6, takeoff, 88
    Division 7, takeoff, 93
    Division 8, takeoff, 100
    Division 9, takeoff, 105-106

Division 10, takeoff, 108
Division 15, takeoff, 122-123
Division 16, takeoff, 130
Sample Project Plan, 40
SDI (Steel Door Institute), 175
SI Metric system, 5
Site work, 47
Specifications, 41, 133
   metric units used in, 134
Special Construction, 113
Square meter and cubic meter costs, 33
Subcontractors, 26-27
System International (SI) Metric, 5-16

decimal prefixes, 5

T
Takeoff, See Quantity takeoff
Thermal and moisture protection, 89-93

W
Western Wood Products Association (WWPA), 84-85
Windows, 95-100
Wood, 83
WWPA (Western Wood Products Association), 175